Pankaj Thakur
Jatinder Kaur
Editors

Thermal Modeling Reimagined

Recent Research Trends and Applications

DOI: https://doi.org/10.52305/VTJO5568

Library of Congress Cataloging-in-Publication Data

ISBN: 979-8-89530-463-1 (Softcover)
ISBN: 979-8-89530-645-1 (eBook)

Published by Nova Science Publishers, Inc. † New York

Contents

Preface

Thermal dynamics and thermoelasticity are pivotal in understanding the behavior of engineering materials and structures under varying temperatures and stresses. This collection brings together cutting-edge research and analytical approaches in these fields, with a special focus on rotating disks and other thermally influenced structures. The chapters compiled in this volume are aimed at advancing the theoretical and practical understanding of thermal stresses, offering new models and computational techniques.

In Chapter 1, the reimagining of thermal dynamics in rotating disks opens the volume with a fresh perspective on the effects of rotation on heat distribution and material deformation. This chapter sets the tone for exploring the intricacies of thermal stresses in rotating systems, laying the groundwork for subsequent discussions.

Chapter 2 introduces a sophisticated mathematical model to capture the interactions of thermal stresses in three-component materials, an essential contribution for those working with complex material systems in industrial and research settings.

Chapter 3 expands the scope by offering a three-dimensional analysis of the thermally stressed state in layered cylindrical shells. This chapter provides valuable insights for engineers and researchers working with multilayered materials and structures, where thermal behavior is critical to design and performance.

In Chapter 4, an in-depth analysis is provided on the effects of circular thickness and bi-linear thermal gradients on the time period of parallelogram-shaped plates. This study is particularly useful for those involved in structural analysis, where thermal and geometric factors must be carefully balanced for optimal performance.

Chapter 5 delves into the unraveling of thermal stress in rotating disks, specifically addressing transversely isotropic materials. The analytical approach presented offers new ways of understanding how these materials react under rotation and thermal stress, which is essential for a variety of mechanical and industrial applications.

The focus shifts to computational techniques in Chapter 6, where the authors tackle a boundary value problem in thermoelastic microelongated solids using the finite difference method. This chapter contributes significantly to the development of numerical methods in solving complex thermoelastic problems.

Finally, Chapter 7 explores the impact of temperature variation on rotating solid disks, with a focus on angular speed, stress, and displacement. This chapter synthesizes much of what has been discussed earlier, providing a comprehensive overview of how temperature changes influence the mechanical properties of rotating systems.

This volume represents a comprehensive and detailed exploration of thermal dynamics in various systems, providing readers with new approaches and models that are both theoretically rigorous and practically applicable. The research presented here will be invaluable to scientists, engineers, and academicians working in the fields of thermoelasticity, materials science, and structural analysis.

Acknowledgments

We would like to express our deepest gratitude to the outstanding authors who have contributed their expertise, insights, and creativity to this book. Your dedication to the craft of writing and your passion for sharing knowledge has enriched this project beyond measure.

To our national authors, thank you for your invaluable contributions.

Your dedication to advancing literature and research efforts within our nation in this particular field has significantly enriched the content of this book.

We extend our heartfelt appreciation to our international authors. Your global perspective and willingness to bridge cultural divides have elevated the content of this book to a new level of excellence. Your willingness to participate in this project, despite geographical boundaries, is a testament to the unifying power of knowledge.

We owe a special debt of gratitude to Nova Science Publishers, Inc., USA, for their unwavering support and commitment to the success of this book. Their expertise in the field of publishing and dedication to the dissemination of knowledge has been invaluable. We are deeply thankful for the collaborative spirit, patience and professionalism they exhibited throughout this process.

This book is the result of countless hours of hard work and collaboration from many talented individuals and we extend our appreciation to all those who have played a role in its creation, from editors and designers to proofreaders and advisors. Thank you all for making this book a reality and for contributing to the greater body of knowledge and literature in our field.

With sincere appreciation,

Dr. Pankaj Thakur
Editor

About the Editors

Dr. Pankaj Thakur is working as an associate professor of the Faculty of Science & Technology, at ICFAI University, Himachal Pradesh in India. He received his PhD in Applied Mathematics from Himachal Pradesh University in 2007. Dr. Thakur's research has been in the area of solid mechanics, elasticity, plasticity, creep, materials and thermal stress. He has authored more than 200 publications and has given numerous presentations at national and international meetings. Dr. Thakur guided 16 PhD candidates, and 30 postgraduate candidates. He completed a major research project funded by the University Grants Commission, India. He is currently the Section Editor-in-Chief of Mathematics/Statistic Journal of Nigerian Society of Physical Science and a member of editorial and reviewer boards of several journals (index in SCOPUS/WOS).

Dr. Jatinder Kaur is currently serving as a professor in the Department of Mathematics at Chandigarh University in Mohali, India. With a rich teaching experience spanning over 15 years. She has taught in both college and university settings. In 2017, she earned her PhD in Applied Mathematics from Punjabi University, Patiala. Her research interests encompass a wide range of topics, including elasticity, plasticity, mechanics, creep, materials, and manifolds. Her contributions to the academic world are evident through her publication of more than 15 research papers in esteemed national and international journals. In addition to her research pursuits, Dr. Kaur plays a pivotal role in academia by guiding MSc. and PhD students in their scholarly journeys. Notably, she has also served as a valued reviewer for numerous prestigious journals, making significant and continuous contributions to the field of research.

Chapter 1

Reimagining Thermal Dynamics in Rotating Disks

Pankaj Thakur*, PhD
Priya Gulial, MSc
and Palvinder Thakur, MSc
Faculty of Science and Technology, ICFAI University Himachal Pradesh, Kalu Jhanda, Himachal Pradesh, India

Abstract

This study presents an analytical exploration of thermal stress in rotating disks crafted from transversely isotropic materials. By unraveling this complex interplay, it sheds light on the nuanced mechanisms governing stress distribution. Insights gleaned promise to enhance the design and resilience of rotating components in various engineering applications.

Keywords: analytical modelling, thermal stress, disk, density, angular speed

Introduction

The study of thermal dynamics in rotating disks is critical for various engineering applications, including aerospace, automotive, and industrial machinery. As these disks operate under mechanical loads and varying temperatures, understanding their thermal behavior becomes essential for

---

* Corresponding Author's Email: pankaj_thakur15@yahoo.co.in

In: Thermal Modeling Reimagined
Editors: Pankaj Thakur and Jatinder Kaur
ISBN: 979-8-89530-463-1

optimizing performance and ensuring safety. Traditional thermal models often fall short in accurately predicting the interplay between material properties, rotational speed, and thermal effects. Recent advancements in computational techniques and material science have opened new avenues for reimagining these thermal dynamics. This research paper explores contemporary trends in thermal modeling, focusing on how innovative approaches can enhance our understanding of heat distribution and stress responses in rotating disks. By integrating multi-physics simulations and experimental data, we aim to develop more robust models that account for different materials and loading conditions. This investigation not only addresses the limitations of existing methods but also proposes novel applications that can significantly impact design and operational efficiency. Through a comprehensive analysis, this paper seeks to provide valuable insights for engineers and researchers, paving the way for future advancements in the field of thermal dynamics in rotating systems. The examination of rotating disks under mechanical and thermal loads has been conducted using both linear and nonlinear methodologies. In linear analyses, researchers predominantly rely on infinitesimal elasticity theory to study isotropic and anisotropic disks with uniform thickness profiles [1-3]. In contrast, nonlinear research has focused on three key dimensions: geometric nonlinearity, material properties, and analytical techniques. While many earlier investigations of rotating disks [4] addressed those with uniform thickness, subsequent studies have highlighted the significance of variable thickness in nonlinear geometries [5-6]. Recent findings [7-9] have shown that stresses in rotating disks, whether solid or annular, are significantly reduced when variable thickness is considered, especially at the same angular velocity. Additionally, while many studies have concentrated solely on mechanical loads, a considerable amount of literature has also explored the effects of thermal loads on disks [10-15]. Accurately determining stress distributions in these rotating structures is crucial for enhancing design and material efficiency in various engineering applications, including rotors, flywheels, turbines, and computer disk drives. For instance, Thakur [16] utilized Seth's transition theory to analyze stresses in thin rotating disks with inclusions. The elastic behavior of isotropic materials has been extensively discussed by Timoshenko and Goodier [17], while Chakrabarty [18] and Heyman [19] have addressed the plastic behavior. Güven [20] tackled the complexities related to rigid inclusions under Tresca's yield conditions, emphasizing that simplistic assumptions about material behavior can lead to unrealistic models, as perfect elasticity and ideal plasticity represent two extremes that cannot be distinctly separated.

Objective of the Study

The objective of this study is to advance the understanding of thermal dynamics in rotating disks by integrating innovative modeling techniques and experimental data. This research aims to explore the effects of varying material properties and thickness profiles on stress distribution and thermal behavior under mechanical and thermal loads. By reexamining traditional assumptions and incorporating nonlinear geometries, the study seeks to provide more accurate predictions of disk performance in real-world applications. Ultimately, the findings aim to enhance design efficiency and material utilization in critical engineering fields such as aerospace, automotive, and industrial machinery.

Governing Equations

A thin annular disc of uniform density, featuring a central bore with radius a and an outer radius b, is analyzed in this study. This disc, made from a material of consistent density, is subjected to edge loading and rotates around a central axis perpendicular to its plane at an angular speed ω. The thickness of the disc is considered constant and sufficiently small, allowing the assumption of a plane stress condition, where the axial stress T_{zz} is negligible. The temperature at the central bore of the disc is denoted as Θ. The rotating disc analyzed in this study is influenced by a temperature gradient and a mechanical load. The inner surface of the disc is assumed to be securely attached to a shaft, ensuring that isothermal conditions are sustained at this boundary. In contrast, the outer surface is subjected to a mechanical load while being maintained at a uniform temperature gradient. Therefore, the boundary conditions for this analysis are defined as follows:

$$u = 0,\ \Theta = \Theta_0 \text{ at } r = a \text{ and } T_{rr} = T_0,\ \Theta = \Theta_0 \text{ at } r = b \tag{1}$$

where $\Theta = \dfrac{\Theta_0 \log(r/b)}{\log(a/b)}$. The stresses are outlined by Thakur [16]:

$$T_{rr} = \frac{2\mu}{n}\left[3 - 2C - \beta^n\left\{1 - C + (2 - C)(P+1)^n + \frac{nC\xi\Theta}{2\mu\beta^n}\right\}\right],$$

$$T_{\theta\theta} = \frac{2\mu}{n}\left[3-2C-\beta^n\left\{2-C+(1-C)(P+1)^n+\frac{nC\xi\Theta}{2\mu\beta^n}\right\}\right],$$

$$T_{r\theta} = T_{\theta z} = T_{zr} = T_{zz} = 0 \tag{2}$$

All equations of equilibrium are satisfied, except for the following:

$$\frac{d}{dr}(rT_{rr}) - T_{\theta\theta} + \rho\omega^2 r^2 = 0, \tag{3}$$

Substituting Equations (1) and (2) into Equation (3) yields a nonlinear differential equation in β:

$$(2-C)n\beta^{n+1}P(P+1)^{n-1}\frac{dP}{d\beta} = \frac{n\rho\omega^2 r^2}{2\mu} - \frac{nC\xi\bar{\Theta}_0}{2\mu} + \beta^n\left[1-(P+1)^n - nP\begin{Bmatrix}1-C\\+(2-C)(P+1)^n\end{Bmatrix}\right] \tag{4}$$

where $\bar{\Theta}_0 = \dfrac{\Theta_0}{\log(a/b)}$. From Equation (4), the turning points of β are $P = -1$ and $\pm\infty$.

Solution

To determine the plastic stress, the transition function is applied through the principal stress (see, Gupta et al. [6], Thakur et al. [5, 7-9, 13-16] and Seth [21]) at the transition point $P \to \pm\infty$. The transition function is defined as follows:

$$\Upsilon = \left(\frac{n}{2\mu}\right)[T_{\theta\theta} - C\xi\Theta] = \left[(3-2C) - \beta^n\left\{2-C+(1-C)(P+1)^n\right\} - \frac{nC\xi\Theta}{\mu}\right] \tag{5}$$

By taking the logarithmic derivative of Equation (5) with respect to r and applying Equation (4), along with the asymptotic value, and integrating, we obtain:

$$\Upsilon = K_1 r^{-1/(2-C)} \tag{6}$$

Equation (5) and (6), become:

$$T_{\theta\theta} = \left(\frac{2\mu}{n}\right) K_1 r^{-1/(2-C)} + \frac{C\xi\Theta_0 \ln(r/b)}{\ln(a/b)} \tag{7}$$

By substituting Equation (7) into Equation (3) and integrating, we obtain:

$$T_{rr} = \left\{\frac{2\mu(2-C)}{1-C}\right\} K_1 r^{-1/(2-C)} + \frac{C\xi\Theta_0 \ln(r/b)}{\ln(a/b)} - \frac{C\xi\Theta_0}{\ln(a/b)} - \frac{\rho\omega^2 r^2}{3} + \frac{K_2}{r} \tag{8}$$

Inserting Equations (7) and (8) into $T_{rr} - T_{\theta\theta}(1 - C/2 - C) = E(2-\beta^2)/2$ given by Thakur [10], we get:

$$\beta = \sqrt{1 - \frac{2(1-C)}{E(2-C)}\left[\frac{\rho\omega^2 r^2}{3} - \frac{K_2}{r} + \frac{\alpha E\Theta_0(2-C)}{\ln(a/b)}\left[1 + \frac{2}{(1-C)}\ln(r/b)\right]\right]} \tag{9}$$

Substituting Equation (19) into $u = r(1-\beta)$, we obtain

$$u = r - r\sqrt{1 - \frac{2\nu}{E}\left[\frac{\rho\omega^2 r^2}{3} - \frac{K_2}{r} + \frac{\alpha E\Theta_0(2-C)}{\ln(a/b)}\left[1 + \frac{2}{(1-C)}\ln(r/b)\right]\right]} \tag{10}$$

where $\nu = 1 - C/2 - C$.

By applying boundary condition (1) to Equations (8) and (10), we obtain:

$$K_1 = \frac{n\nu}{2\mu b^{\nu}}\left[bT_0 + \frac{\rho\omega^2(b^3 - a^3)}{3}\right] + \frac{\alpha E\Theta_0(1-C)(b-a)}{2\mu \ln(a/b) b^{\nu}} - \frac{\alpha E\Theta_0 na}{\mu b^{\nu}}$$

and

$$K_2 = \frac{\rho\omega^2 a^3}{3} + \frac{\alpha E\Theta_0(2-C)}{\ln(a/b)}\left[1 - \frac{2}{(1-C)}\ln\left(\frac{r}{b}\right)\right].$$

Substituting the values of constants K_1 and K_2 from Equations (7), (8), and (10), respectively, we obtain:

$$T_{\theta\theta} = \left\{ \begin{array}{l} \dfrac{\rho\omega^2 \left(b^3 - a^3\right)\nu}{3r}\left(\dfrac{r}{b}\right)^{\nu} + \nu T_0 \left(\dfrac{r}{b}\right)^{\nu-1} - 2\alpha E\Theta_0 \left(\dfrac{a}{r}\right)\left(\dfrac{r}{b}\right)^{\nu} \\ +\alpha E\Theta_0 (2-C)\left[\dfrac{(1-C)(b-a)}{r(2-C)\ln(a/b)}\left(\dfrac{r}{b}\right)^{\nu} - \dfrac{\ln(r/b)}{\ln(a/b)}\right] \end{array} \right\} \tag{11}$$

$$T_{rr} = \left\{ \begin{array}{l} \dfrac{\rho\omega^2}{3r}\left[\left(\dfrac{r}{b}\right)^{\nu}\left(b^3 - a^3\right) - r^3 + a^3\right] + \left(\dfrac{r}{b}\right)^{\nu-1} T_0 + \dfrac{2\alpha E\Theta_0}{\nu}\left[\dfrac{a}{r} - \dfrac{a}{r}\left(\dfrac{r}{b}\right)^{\nu}\right] \\ + \dfrac{\alpha E\Theta_0 (2-C)}{\ln(a/b)}\left[\ln(r/b) + \dfrac{(b-a)}{r}\left(\dfrac{r}{b}\right)^{\nu} + \dfrac{a}{r} - 1\right] \end{array} \right\} \tag{12}$$

$$u = r - r\sqrt{1 - \frac{2\nu}{E}\left[\frac{\rho\omega^2\left(r^3 - a^3\right)}{3r} + \frac{\alpha E\Theta_0 (2-C)(r-a)}{r\ln(a/b)} + \frac{2(2-C)\alpha E\Theta_0}{(1-C)}\left[\frac{\ln(r/b)}{\ln(a/b)} - \frac{a}{r}\right]\right]} \tag{13}$$

$$\left|T_{rr} - T_{\theta\theta}\right| = \left| \begin{array}{l} \dfrac{\rho\omega^2}{3r}\left[\left(\dfrac{r}{b}\right)^{\nu}\left(b^3 - a^3\right)(1-\nu) - r^3 + a^3\right] + \left(\dfrac{r}{b}\right)^{\nu-1}(1-\nu)T_0 + \\ \alpha E\Theta_0\left[2\left(\dfrac{a}{r}\right)\dfrac{1}{\nu} - \dfrac{2a}{(1-C)r}\left(\dfrac{r}{b}\right)^{\nu} + \dfrac{(b-a)}{r.\ln(a/b)}\left(\dfrac{r}{b}\right)^{\nu} + \dfrac{1}{\log(a/b)(1-\nu)}\left(\dfrac{a-r}{r}\right)\right] \end{array} \right| \tag{14}$$

It can be observed that $\left|T_{rr} - T_{\theta\theta}\right|$ reaches its maximum at the inner surface. Therefore, yielding will occur at the internal surface of the disc, and Equation (14) provides:

$$\left|T_{rr} - T_{\theta\theta}\right|_{r=a} = \left| \begin{array}{l} \dfrac{\rho\omega^2\left(b^3 - a^3\right)(1-\nu)}{3a}\left(\dfrac{a}{b}\right)^{\nu} + \left(\dfrac{a}{b}\right)^{\nu-1}(1-\nu)T_0 \\ +\alpha E\Theta_0\left[\dfrac{(b-a)}{a.\ln(a/b)}\left(\dfrac{a}{b}\right)^{\nu} - \dfrac{2}{(1-C)}\left(\dfrac{a}{b}\right)^{\nu} + \dfrac{2}{\nu}\right] \end{array} \right| = Y(say) \tag{15}$$

The angular speed required for initial yielding is derived from Equation (15):

$$\Omega_i^2 = \frac{\rho\omega_i^2 b^2}{Y} = \left| \frac{3ab^2\left[1-\left(\frac{T_0}{Y}\right)(1-\nu)\left(\frac{a}{b}\right)^{\nu-1}\right]}{\left(\frac{a}{b}\right)^{\nu}\left(b^3-a^3\right)(1-\nu)} \right| - \left| \left(\frac{3}{(1-\nu)\left(1-a^3/b^3\right)}\right)\left(\frac{\alpha E\Theta_0}{Y}\right) \left[\frac{(1-a/b)}{\ln(a/b)} - \frac{2}{\nu}(1-\nu)\left(\frac{a}{b}\right) + \frac{2}{\nu}\left(\frac{a}{b}\right)^{1-\nu}\right] \right| \tag{16}$$

Equations (11)–(13) and (16) in non-dimensional form become:

$$\sigma_\theta = \frac{\Omega_i^2}{3R}\nu\left(1-R_0^3\right)R^\nu + \nu R^{\nu-1}\sigma_0 + \frac{\Theta_1}{1-\nu}\left[\frac{(1-R_0)\nu}{R.\ln R_0}R^\nu + \frac{\ln R}{\ln R_0} - 2R_0(1-\nu)R^{\nu-1}\right] \tag{17}$$

$$\sigma_r = \left\{ \begin{aligned} &\sigma_0 R^{\nu-1} + \frac{\Omega_i^2}{3R}\left[R^\nu\left(1-R_0^3\right) - R^3 + R_0^3\right] + \frac{2\Theta_1}{\nu}\left[\frac{R_0}{R} - \frac{R_0}{R}R^\nu\right] \\ &+ \frac{\Theta_1}{(1-\nu)\ln R_0}\left[\ln R + \frac{(1-R_0)}{R}R^\nu + \frac{R_0}{R} - 1\right] \end{aligned} \right\} \tag{18}$$

$$U = R - R\left\{1 - 2\nu H\left[\frac{\Omega_i^2}{3R}\left(R^3 - R_0^3\right) + \frac{2\Theta_1(R-R_0)}{R.\ln R_0(1-\nu)} + \frac{2\Theta_1}{\nu}\left[\frac{\ln R}{\ln R_0} - \frac{R_0}{R}\right]\right]\right\}^{1/2} \tag{19}$$

and

$$\Omega_i^2 = \left| \frac{3}{\left(1-R_0^3\right)(1-\nu)}R_0^{\nu-1} - \frac{3\sigma_0}{\left(1-R_0^3\right)} \right| - \left| \left(\frac{3\Theta_1}{(1-\nu)\left(1-R_0^3\right)}\right)\left[\frac{(1-R_0)}{\ln R_0} - \frac{2}{\nu}R_0^{1-\nu} + \frac{2(1-\nu)}{\nu}R_0\right] \right| \tag{20}$$

The angular speed of the rotating disc reaches full plasticity at the external surface, and Equation (14) becomes:

$$\left|T_{rr} - T_{\theta\theta}\right|_{r=b} = \left| \frac{1}{2}T_0 - \frac{\rho\omega^2\left(b^3-a^3\right)}{6b} + \alpha E\Theta_0\left[\frac{(b-a)}{b.\ln(a/b)} - 2\left(\frac{a}{b}\right)\right] \right| \equiv Y^*(say) \tag{21}$$

The angular speed required for the disc to become fully plastic is given by Equation (21):

$$\Omega_f^2 = \frac{\rho\omega_f^2 b^2}{Y^*} = \left|\frac{6}{1-\left(a^3/b^3\right)}\cdot\left(\frac{\sigma_0}{2}-1\right)\right| - \left|\frac{6}{1-\left(a^3/b^3\right)}\left(\frac{\alpha E\Theta}{Y^*}\right)\left[\frac{\left(1-a/b\right)}{\ln(a/b)}-2\left(\frac{a}{b}\right)\right]\right| \tag{22}$$

Equations (17)–(19) and (22) for the fully plastic state become:

$$\sigma_\theta = \frac{1}{2\sqrt{R}}\left[\sigma_0 + \frac{\Omega_f^2}{3}\left(1-R_0^3\right)\right] + 2\Theta_1\left[\frac{\left(1-R_0\right)}{2R.\ln R_0}R^{1/2} + \frac{\ln R}{\ln R_0} - \frac{R_0}{\sqrt{R}}\right] \tag{23}$$

$$\sigma_r = \left\{\begin{array}{l} \dfrac{\sigma_0}{\sqrt{R}} + \dfrac{\Omega_f^2}{3R}\left[\sqrt{R}\left(1-R_0^3\right) - R^3 + R_0^3\right] + 4\Theta_1\left[\dfrac{R_0}{R} - \dfrac{R_0}{\sqrt{R}}\right] \\ + \dfrac{2\Theta_1}{\ln R_0}\left[\ln R + \dfrac{\left(1-R_0\right)}{R}R^{1/2} + \dfrac{R}{R_0} - 1\right] \end{array}\right\} \tag{24}$$

$$U = R - R\left\{1 - H\left[\frac{\Omega_i^2}{3R}\left(R^3 - R_0^3\right) + \frac{4\Theta_1\left(R-R_0\right)}{R.\ln R_0} + 4\Theta_1\left[\frac{\ln R}{\ln R_0} - \frac{R_0}{R}\right]\right]\right\}^{1/2} \tag{25}$$

$$\Omega_f^2 = \left|\frac{6}{\left(1-R_0^3\right)}\cdot\left(\frac{\sigma_0}{2}-1\right)\right| - \left|\frac{6\Theta_1}{\left(1-R_0^3\right)}\left[\frac{\left(1-R_0\right)}{\ln R_0} - 2R_0\right]\right| \tag{26}$$

The results obtained from Equations (17)–(26) are consistent with those presented by Thakur [16], assuming that load and thermal conditions are neglected.

Results and Discussion

For calculating the thermal stresses, angular speed and displacement based on the above analysis, numerically the following values have been taken: $\nu = 0.5$ (i.e., rubber material), 0.333 (i.e., copper), $E/Y = 1$, $\Theta_1 = 0$ and 1.

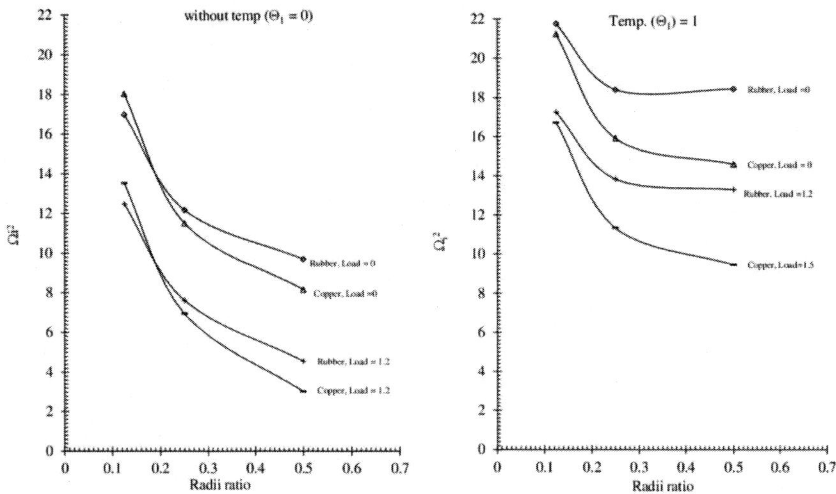

Figure 1. Angular speed versus Radii ratio.

In Figure 1, the curves represent the relationship between angular speed Ω_i^2 and the radial ratio R_0 for both copper and rubber, with and without temperature effects. Across both graphs, Ω_i^2 consistently decreases as the radial ratio increases, but copper shows a steeper decline compared to rubber. Without temperature, Ω_i^2 is generally lower for both materials, whereas with temperature effects, Ω_i^2 starts at a higher value and decreases more gradually, particularly for copper. Copper is more sensitive to variations in radial ratio, temperature, and load, exhibiting a more pronounced reduction in Ω_i^2 under these conditions. In contrast, rubber experiences a more gradual decrease in Ω_i^2, demonstrating less sensitivity to these factors. The impact of load is also significant, with higher loads causing a greater reduction in Ω_i^2, especially for copper, an effect that becomes even more pronounced when temperature is considered. Overall, copper is more responsive to changes in radial ratio and external factors like temperature and load, while rubber shows a more stable behavior.

Conclusion

The results of this chapter are as follows:

- Without temperature, Ω_i^2 is generally lower for both materials. With temperature effects, Ω_i^2 starts higher but decreases more gradually, especially for copper.
- Copper is more responsive to both radial ratio and external conditions (temperature and load) compared to rubber, which maintains a more stable behavior.

Disclaimer

None.

References

[1] Johnson W. & Mellor P. B. (1978). *Engineering plasticity*, London: Von Nastrand Reinhold.

[2] Altenbach H. & Skrzypek J. J. (1999). *Creep and damage in materials and structures*, Springer Verlag, Berlin.

[3] Swainger K. H. (1956). *Analysis of Deformation*, Chapman & Hall, London; Macmillan, USA, Vol. III, Fluidity, 67-68.

[4] Ghose N. C. (1975). Thermal effect on the transverse vibration of spinning disk of variable thickness, *J. Appl. Mech.*, 42, 358-362.

[5] Thakur P., Kaur J. & Singh S. B. (2016). Thermal creep transition stresses and strain rates in a circular disc with shaft having variable density, *Eng. Comput.*, 33(3), 698-712.

[6] Gupta S. K. & Thakur P. (2007). Thermo elastic-plastic transition in a thin rotating disc with inclusion, *Thermal Science*,11(1),103-118.

[7] Thakur P., Kaur J. & Sharma P. L. (2024). *New Research on Thermal Stresses*, Nova Science Publisher USA, 1–146.

[8] Thakur P. & Sethi M. (2020). Creep deformation and stress analysis in a transversely material disc subjected to rigid shaft, *Mathematics and Mechanics of Solids*, 25(1), 17-25.

[9] Sethi M. & Thakur P. (2020). Elastoplastic deformation in an isotropic material disk with shaft subjected to load and variable density, *Journal of Rubber Research*, 23(2), 69-78.

[10] Murakami S. & Konishi K. (1982). An elastic-plastic constitutive equation for transversely isotropic materials and its application to the bending of perforated circular plates, *International Journal of Mechanical Sciences*,24 (12), 763-775.

[11] Temesgen A. G., Singh S. B. & Thakur P. (2020). Modeling of creep deformation of a transversely isotropic rotating disc with a shaft having variable density and subjected to a thermal gradient, *Thermal Science and Engineering Progress*, 20 (100745).

[12] Temesgen A. G., Singh S. B. & Thakur P. (2021). Elastoplastic analysis in functionally graded thick-walled rotating transversely isotropic cylinder under a radial temperature gradient and uniform pressure, *Mathematics and Mechanics of Solids*, 26(1),5 -17.

[13] Thakur P., Kumar N. & Sethi M. (2021). Elastic-plastic stresses in a rotating disc of transversely isotropic material fitted with a shaft and subjected to thermal gradient, *Meccanica*, 56(5),1165-1175.

[14] Thakur P., Sethi M., Gupta N., Gupta K. & Bhardwaj R. K. (2021). Thermal stress analysis in a hemispherical shell made of transversely isotropic materials under pressure and thermo-mechanical loads, *ZAMM*, 101(12), e202100208.

[15] Thakur P., Sethi M., Kumar N., Gupta N., Gupta K. & Bhardwaj R. K. (2022). Stress analysis in an isotropic hyperbolic rotating disk fitted with rigid shaft, *Zeitschrift für angewandte Mathematik und Physik*., 73(1), article id 23,1-11.

[16] Thakur P. (2010). Elastic-plastic transition stresses in a thin rotating disc with rigid inclusion by infinitesimal deformation under steady state Temperature, *Thermal Science*, 14(1), 209-219.

[17] Timoshenko, S. P. & Goodier, J. N. (1970). *Theory of Elasticity*, McGraw-Hill, New York.

[18] Chakrabarty, J. (1987). *Theory of Plasticity*, McGraw- Hill, New York, USA.

[19] Heyman, J. (1958). Plastic Design of Rotating Discs, *Proc. Inst. Mech. Engrs*., 172(3), 531-546.

[20] Güven U. (1999). Elastic - plastic rotating disk with rigid inclusion, *Mech. Struct. & Mach.*, 27, 117-128.

[21] Seth B. R. (1966). Measure-concept in mechanics, *Int. J Non-linear Mech.*, 1(1), 35-40.

Chapter 2

Mathematical Model of Interactions of Thermal Stresses in Three-Component Materials

Ladislav Ceniga[*]**, DSc**
Institute of Materials Research, Slovak Academy of Sciences, Kosice, Slovak Republic

Abstract

This chapter presents a mathematical model for the interaction of thermal stresses in a material system corresponding to accurate three-component materials. Analytical modelling is based on the fundamental equations of solid elastic continuum mechanics. The thermal stresses, which arise during the cooling process, result from the different thermal expansion coefficients of the material components. The model system consists of isotropic spherical particles with an isotropic spherical envelope on their surface, periodically distributed within an isotropic matrix. The matrix is imaginatively divided into identical cubic cells containing a central particle. The dimensions of the cubic cells, which are similar to the inter-particle distance, along with the radii of the particles/envelope and the particle volume fraction, are microstructural parameters of the three-component materials. The mathematical model, which incorporates these microstructural parameters, is developed concerning the cubic cell and is derived using the superposition method of continuum mechanics to determine the mutual interactions of thermal stress fields around neighbouring material components. Finally, the mathematical model for a thermal-stress-induced crack is also developed.

---

[*] Corresponding Author's Email: lceniga@yahoo.com

In: Thermal Modeling Reimagined
Editors: Pankaj Thakur and Jatinder Kaur
ISBN: 979-8-89530-463-1

Keywords: mathematical model, thermal stress, stress interaction, three-component material, crack formation, limit state

Introduction

This chapter presents an original mathematical model for the interaction of thermal stresses in the isotropic components of three-component materials. These thermal stresses, which arise during the cooling process, occur below the relaxation temperature of the material. The mathematical model is derived using the superposition method of solid continuum mechanics, considering mechanical loading, mechanical bonds, and the shape of the solid continuum. This method determines the mutual interactions of the thermal stress fields around neighboring material components. As is typical in solid continuum mechanics, these superposition models are developed through a standard procedure, which includes:

- Defining the model material system to correspond to accurate three-component materials,
- Defining a coordinate system to correspond to the model material system,
- Analyzing the cause of the thermal stresses,
- Defining the fundamental equations of solid continuum mechanics, which are transformed into partial differential equations,
- Defining the total potential energy of the model material system concerning different mathematical solutions,
- Analyzing mathematical boundary conditions,
- Applying suitable mathematical procedures to the partial differential equations to obtain solutions,
- Conducting a mathematical analysis of crack formation in the model material system.

Model Material System

In the context of mathematical modeling, accurate three-component materials with finite dimensions consist of isotropic precipitates surrounded by an isotropic continuous component (envelope) on their surface, distributed within isotropic material grains. The multi-particle-envelope-matrix model system with infinite dimensions, illustrated in Figure 1, comprises an endless matrix and spherical particles encased in a spherical envelope, as described by Ceniga [1]. The precipitates, continuous components, and material grains—representing the components of an accurate three-component material—correspond to the spherical particles, spherical envelopes, and infinite matrix of the multi-particle-envelope-matrix model system, respectively. This model material system is conceptually divided into identical cubic cells with a dimension *d*. Each cell contains a central spherical particle with a spherical envelope, where R_1 is the radius of the particle, and R_1 and R_2 are the radii of the envelope.

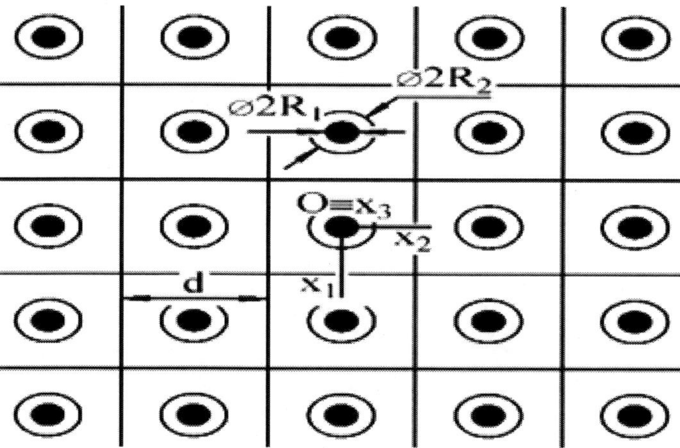

Figure 1. The multi-particle-envelope-matrix system is conceptually divided into identical cubic cells with dimension *d* , each containing a central spherical particle. The spherical particles, each with a spherical envelope on its surface, are periodically distributed within the infinite matrix. In this system, R_1 is the radius of the particles, and R_1 and R_2 are the radii of the envelope. The matrix extends infinitely along the x_1, x_2, and x_3 axes of the Cartesian coordinate system ($Ox_1x_2x_3$), where O is the center of the spherical particles.

The cubic cell represents a portion of the model material system corresponding to a single spherical particle. The matrix extends infinitely along the x_1, x_2, and x_3 axes of the Cartesian coordinate system ($Ox_1x_2x_3$), where O denotes the center of the spherical particle. The model material system is depicted in Figure 1 within the x_1x_2 plane of the Cartesian coordinate system ($Ox_1x_2x_3$). Due to the infinite extent of the matrix, the same configuration applies to the x_1x_3 and x_2x_3 planes as well. Regarding the particle volume ($V_p = 4\pi R_1^3/3$) and the cell volume ($V_c = d^3$), the particle volume fraction ($v_p = V_p/V_c$) is derived as:

$$v_p = \frac{V_p}{V_c} = \frac{4\pi}{3}\left(\frac{R_1}{d}\right)^3 \in \left(0, v_{p\max}\right\rangle, v_{p\max} = \frac{\pi}{6}\left(\frac{R_1}{R_2}\right)^3,$$

$$d = R_1\left(\frac{4\pi}{3v_p}\right)^{1/3} \tag{1}$$

The maximum value of v_p, denoted as $v_{p\ max}$, is obtained under the condition $d=2R_2$. The parameters v_p, d, R_1, and R_2 represent structural characteristics of a real three-component material, which are determined through experimental and computational techniques.

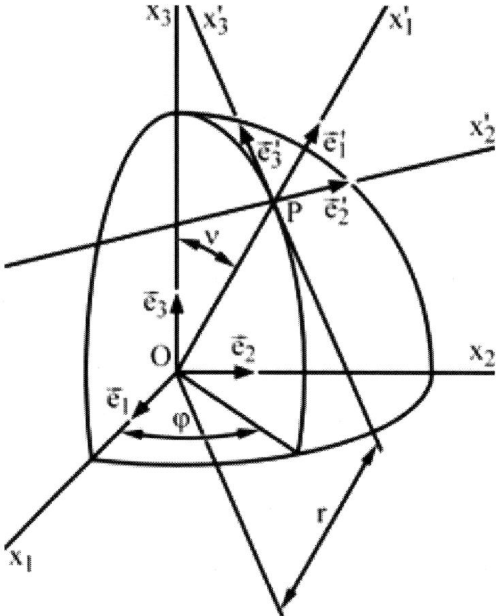

Figure 2. The arbitrary point P is defined in spherical coordinates (r,ϕ,ν) with respect to the Cartesian coordinate system ($Ox_1x_2x_3$) (see Figure 1). Here, $\bar{e}_i$ (i = 1,2,3) represents the unit vector along the axis x_i. The axes x'_1 and x'_2, which lie parallel to the x_1x_2 plane, and x'_3 represent the radial and tangential directions, respectively, where $\bar{e}_i'$ denotes a unit vector along the axis x'_1.

Analytical and computational models of phenomena in infinite periodic material systems are developed within identical, representative cells, as described by Mura [2]. Owing to the infinity and periodicity of the system, the analytical and computational results obtained for a specific cell are applicable to any cell. Furthermore, infinite matrices are assumed for mathematical simplicity, facilitating analytical and computational solutions. These solutions are considered sufficiently accurate for relatively small material components compared to macroscopic material samples, structural elements, and similar systems. Despite this simplification, the solutions are deemed acceptable.

Thermal stresses are evaluated at the arbitrary point *P* along the axes x'_1, x'_2, and x'_3 of the Cartesian coordinate system ($Px'_1x'_2x'_3$), with corresponding unit vectors $\bar{e}_1$, $\bar{e}_2$, and $\bar{e}_3$ (see Figure 2). The point P is defined by spherical coordinates (r,ϕ,ν) (see Figure 2). The axes x'_1, x'_2, and x'_3, with unit vectors $\bar{e}_1$, $\bar{e}_2$, $\bar{e}_3$, represent radial and tangential directions with respect to the spherical surface of radius $r = |OP|$, respectively, where $x'_2 || x_1x_2$. Spherical coordinates (r,ϕ,ν) are used because of the spherical geometry of the particles and their envelope, as described by Ceniga [3].

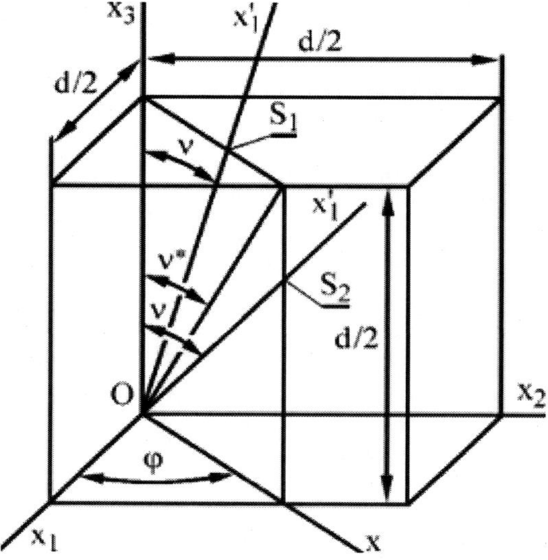

Figure 3. One-eighth of the cubic cell is shown in Figure 1. The points S_1 and S_2 represent the intersections of the axis x'_1 with the surfaces 3657 and 1456 for $\nu \in \langle 0, \nu^* \rangle$ and $\nu \in \langle \nu^*, \pi/2 \rangle$, respectively. The axis x'_1 represents a radial direction (see Figure 2), which is defined by the angles $\varphi = \angle(xx'_1) \in \langle 0, \pi/2 \rangle$, and $\nu \in \langle 0, \pi/2 \rangle, x \in x_1 x_2$.

The thermal stresses are determined within the cubic cell. The multi-particle envelope-matrix system shown in Figure 1 is symmetric. Accordingly, the thermal stresses can be adequately determined within one-eighth of the cubic cell (see Figure 3), i.e., for $\varphi\in\langle 0,\pi/2\rangle$ and $\nu\in\langle 0,\pi/2\rangle$. The intervals $r\in\langle 0,R_1\rangle$, $r\in\langle R_1, R_2\rangle$, and $r\in\langle R_2,r_s\rangle$ correspond to the spherical particle, the spherical envelope, and the cell matrix, respectively. The distance r_s is derived as described by Ceniga [4].

$$r_S = OS_1 = \frac{d}{2\left(c_\varphi + \tan\nu\right)\cos\nu}\left[\frac{c_\varphi\left(1+\cos^2\varphi\right)-\cos\varphi}{\cos^2\varphi} + \tan\nu\right],\quad \nu\in\left\langle 0,\nu^*\right\rangle;$$

$$r_S = OS_2 = \frac{d}{2c_\varphi \sin\nu},\quad \nu\in\left\langle \nu^*,\frac{\pi}{2}\right\rangle,$$

$$\nu^* = \arctan\left(\frac{O8}{O3}\right) = \arctan\left(\frac{1}{c_\varphi}\right);\ c_\varphi = \cos\varphi,\ \ \varphi\in\left\langle 0,\frac{\pi}{4}\right\rangle;\ c_\varphi = \sin\varphi,\ \ \varphi\in\left\langle \frac{\pi}{4},\frac{\pi}{2}\right\rangle, \tag{2}$$

Solid Continuum Mechanics

The model material system shown in Figure 1 is symmetric. The axis x'_1 (see Figure 2) represents the normal to the particle-envelope and matrix-envelope boundaries. Due to this symmetry, any point along x'_1 undergoes a radial displacement u'_1, while $u'_2 = 0$ and $u'_3 = 0$ represent the tangential displacements along the axes x'_2 and x'_3, respectively. Cauchy's equations define the relationships between displacements and strains in a solid continuum. The radial strain ε'_{11}, the tangential strains ε'_{22}, $\alpha\nu\delta$ ε'_{33}, as well as the shear strains ε'_{12} and ε'_{13}, are derived as described by Ceniga [4].

$$\varepsilon'_{11} = \frac{\partial u'_1}{\partial r}, \tag{3}$$

$$\varepsilon'_{22} = \varepsilon'_{33} = \frac{u'_1}{r}, \tag{4}$$

$$\varepsilon_{12}' = \frac{1}{r}\frac{\partial u_1'}{\partial \varphi}, \tag{5}$$

$$\varepsilon_{13}' = \frac{1}{r}\frac{\partial u_1'}{\partial \nu}. \tag{6}$$

In the case of shear strain and stress, ε'_{23} and σ'_{23}, we have $\varepsilon'_{23} = 0$ and $\sigma'_{23} = 0$, respectively, due to $u'_2 = 0$ and $u'_3 = 0$. Regarding Equations (3)–(6), Hooke's law for an isotropic elastic solid continuum is expressed in the following forms for the radial stress σ'_{11}, the tangential stresses σ'_{22} and σ'_{33}, as well as the shear stresses $\sigma'_{12} = \sigma'_{21}$ and $\sigma'_{13} = \sigma'_{31}$:

$$\sigma_{11}' = (c_1 + c_2)\frac{\partial u_1'}{\partial r} - 2c_2\frac{u_1'}{r}, \tag{7}$$

$$\sigma_{22}' = \sigma_{33}' = -c_2\frac{\partial u_1'}{\partial r} + c_1\frac{u_1'}{r}, \tag{8}$$

$$\sigma_{12}' = \frac{1}{s_{44}r}\frac{\partial u_1'}{\partial \varphi}, \tag{9}$$

$$\sigma_{13}' = \frac{1}{s_{44}r}\frac{\partial u_1'}{\partial \nu}, \tag{10}$$

The coefficient c_i ($i = 1, 2, 3$) is derived as follows:

$$c_1 = \frac{E}{(1+\mu)(1-2\mu)}, \quad c_2 = -\frac{\mu E}{(1+\mu)(1-2\mu)}, \quad c_3 = -4(1-\mu) < 0, \tag{11}$$

Here, E and μ represent Young's modulus and Poisson's ratio, respectively. For an isotropic elastic solid continuum, $\mu=0.25$ (Brdicka et al. [5]). For a real isotropic material, $\mu < 0.5$ (Skocovsky et al., 1996). Due to $\mu<0.5$, we find that $c_3<0$.

The equilibrium equations for the radial stress σ'_{11}, the tangential stresses σ'_{22} and σ'_{33}, as well as the shear stresses $\sigma'_{12} = \sigma'_{21}$ and $\sigma'_{13} = \sigma'_{31}$, are derived as described by Ceniga [6].

$$2\sigma_{11}^{'} - \sigma_{22}^{'} - \sigma_{33}^{'} + \frac{\partial \sigma_{11}^{'}}{\partial r} + \frac{\partial \sigma_{12}^{'}}{\partial \varphi} + \frac{\partial \sigma_{13}^{'}}{\partial \nu} = 0 , \tag{12}$$

$$\frac{\partial \sigma_{22}^{'}}{\partial \varphi} + 3\sigma_{12}^{'} + \frac{\partial \sigma_{12}^{'}}{\partial r} = 0 , \tag{13}$$

$$\frac{\partial \sigma_{33}^{'}}{\partial \nu} + 3\sigma_{13}^{'} + \frac{\partial \sigma_{13}^{'}}{\partial r} = 0 . \tag{14}$$

By substituting Equations (7)–(10) into Equation (12) and into the sum ∂Equation (13)/∂φ+∂Equation(14)/∂ν, the equilibrium Equations (12)–(14) are transformed into the following forms:

$$r^2 \frac{\partial^2 u_1^{'}}{\partial r^2} + 2r \frac{\partial u_1^{'}}{\partial r} - 2u_1^{'} + \frac{U_1^{'}}{s_{44}\left(c_1 + c_2\right)} = 0 , \tag{15}$$

$$r \frac{\partial U_1^{'}}{\partial r} = c_3 U_1^{'} , \tag{16}$$

where the function U'_1 is derived as follows:

$$U_1^{'} = \frac{\partial^2 u_1^{'}}{\partial^2 \varphi} + \frac{\partial^2 u_1^{'}}{\partial^2 \nu} . \tag{17}$$

The elastic energy density w_q, which is accumulated at an arbitrary point in the spherical particle (q = p), the envelope (q=e), and the cell-matrix (q=m), is derived as shown by Brdicka et al. [5] and Ceniga [6]:

$$w_q = \frac{c_{1q} + c_{2q}}{2}\left(\frac{\partial u_{1q}^{'}}{\partial r}\right)^2 - \frac{2c_{2q}u_{1q}^{'}}{r}\frac{\partial u_{1q}^{'}}{\partial r} + \frac{c_{1q}\left(u_{1q}^{'}\right)^2}{r^2} + \frac{1}{s_{44}r^2}\left[\left(\frac{\partial u_{1q}^{'}}{\partial \varphi}\right)^2 + \left(\frac{\partial u_{1q}^{'}}{\partial \nu}\right)^2\right], \quad q = p, e, m \tag{18}$$

In the case of spherical coordinates (r,ϕ,ν) (see Figure 2), the elastic energy W_q ($q = p,e,m$), which is accumulated in the volume V_q, along with the energy W_c, which is accumulated in the cubic cell, is derived as follows:

$$W_q = \int_{Vq} w_q \, dV = 8 \int_0^{\pi/2} \int_0^{\pi/2} \int_{r1}^{r2} w_q r^2 \, dr \, d\varphi \, dv, \quad q = p, e, m \quad , \tag{19}$$

$$W_c = W_p + W_e + W_m , \tag{20}$$

where $dV = r^2 dr\, d\varphi dv$, and the limits are $r_1 = 0$, $r_2 = R_1$ for q = p; $r_1 = R_1$,$r_2=R_2$ for q = e; and $r_1 = R_2$, $r_2 = r_S$ for q=m. Different mathematical solutions determine the thermal stresses in the cell matrix. As analyzed in Ceniga [7], this combination of mathematical solutions yields a minimum value for the total potential energy $W_t = W_c$.

Reason for Thermal Stresses

The thermal stresses arise from the conditions $\alpha_p \neq \alpha_e \neq \alpha_m$, $\alpha_p = \alpha_e \neq \alpha_m$ and $\alpha_p \neq \alpha_e = \alpha_m$, where α_p,α_e and ,α_m are the thermal expansion coefficients of the spherical particle, envelope, and matrix, respectively.

The thermal stresses originate at $T \in \langle T_f, T_r \rangle$, where T_f is the final temperature of a cooling process. The relaxation temperature T_r is defined by the formula $T_r = (0.3\text{-}0.4) \times T_m$ (Skocovsky et al. [8]), where TmT_mTm is the melting temperature of the multi-particle-envelope-matrix system (see Figure 1). If $T > T_r$, the thermal stresses are relaxed by thermally activated processes (Skocovsky et al. [8]). Consequently, the coefficient β_q ($q = p$,e, m) is derived as:

$$\beta_q = \int_{Tf}^{Tr} \alpha_q \, dT, \quad q = p,e,m . \tag{21}$$

As mentioned above, the model system in Figure 1 is symmetric. Additionally, about the spherical components, the thermal stresses in this system are a consequence of the radial stresses p_1 and p_2, which act on the particle-envelope and matrix-envelope boundaries, i.e., for $r = R_1$ and $r = R_2$, respectively. The radial stresses p_1 and p_2 are a reason of the radial displacements $(u'_{1p})_{r=R1}$, $(u'_{1e})_{r=R1}$ and $(u'_{1e})_{r=R2}$, $(u'_{1m})_{r=R2}$, respectively. If $\beta_p \neq \beta_e \neq \beta_m$, then we get:

As mentioned above, the model system in Figure 1 is symmetric. Additionally, with respect to the spherical components, the thermal stresses in this system result from the radial stresses p_1 and p_2, which act on the particle-envelope and matrix-envelope boundaries, i.e., at r $r = R_1$ and $r = R_2$, respectively. The radial stresses p_1 and p_2, cause the radial displacements (u'_{1p}) at r = R_1, (u'_{1e}) at r = R$_1$, and (u'_{1e}) at $r = R_2$, (u'_{1m}) at $r = R_2$, respectively. If $\beta_p \neq \beta_e \neq \beta_m$, then we obtain:

$$\left(u'_{1p}\right)_{r=R1} + \left(u'_{1e}\right)_{r=R1} = R_1\left(\beta_p - \beta_e\right), \tag{22}$$

$$\left(u'_{1e}\right)_{r=R2} + \left(u'_{1m}\right)_{r=R2} = R_2\left(\beta_e - \beta_m\right), \tag{23}$$

With regard to (u'_{1p}) at r=R1 = - $R_1\, p_1\, \rho_p$, (u'_{1e}) at r =R$_i$ = - R_i ($p_1\, \rho_{1ie} + p_2\, \rho_{2ie}$) (i=1,2), and (u'_{1m}) at r = R$_2$ = - $R_2\, p_2\, \rho_m$, the radial stresses p_1 and p_2 are derived as:

$$p_1 = \frac{\left(\rho_m + \rho_{22e}\right)\left(\beta_e - \beta_p\right) + \rho_{21e}\left(\beta_e - \beta_m\right)}{\left(\rho_p + \rho_{11e}\right)\left(\rho_m + \rho_{22e}\right) - \rho_{12e}\rho_{21e}}, \tag{24}$$

$$p_2 = \frac{\left(\rho_p + \rho_{11e}\right)\left(\beta_m - \beta_e\right) + \rho_{21e}\left(\beta_p - \beta_e\right)}{\left(\rho_p + \rho_{11e}\right)\left(\rho_m + \rho_{22e}\right) - \rho_{12e}\rho_{21e}}, \tag{25}$$

If $\beta_p \neq \beta_e = \beta_m$, the condition (22) is considered. With regard to $(u'_{1p})_{r=R1}$= - $R_1\, p_1\, \rho_p$, $(u'_{1e})_{r=Ri}$ = - $R_i\, p_1\, \rho_{1me}$, the radial stress p_1 is derived as:

$$p_1 = \frac{\beta_e - \beta_p}{\rho_p + \rho_{1me}}, \tag{26}$$

If $\beta_p = \beta_e \neq \beta_m$, the condition (23) is considered. With regard to $(u'_{1e})_{r=R2}$= - $R_2\, p_2\, \rho_{2pe}$, $(u'_{1m})_{r=Ri}$ = - $R_2\, p_2\, \rho_m$, the radial stress p_2 is derived as:

$$p_2 = \frac{\beta_m - \beta_e}{\rho_m + \rho_{2pe}}, \tag{27}$$

Mathematical Boundary Conditions

Spherical Particle

The absolute value $|u'_{1p}|$ is required to represent an increasing function of $r \in \langle 0, R_1 \rangle$ with a maximum value for $r = R_1$. Additionally, the conditions $(u'_{1p})_{r=0} \neq \pm\infty, (\varepsilon'_{ijp})_{r=0} \neq \pm\infty, (\sigma'_{ijp})_{r=0} \neq \pm\infty (i,j = 1,2,3)$ must be valid. If $\beta_p \neq \beta_e \neq \beta_m$ or $\beta_p \neq \beta_e = \beta_m$, then the mathematical boundary conditions for the radial stress σ'_{11p} and the radial displacement u'_{1p} are derived as:

$$\left(\sigma'_{11p}\right)_{r=R1} = -p_1, \tag{28}$$

$$\left(u'_{1p}\right)_{r=0} = 0, \tag{29}$$

where p_1 is given by Equations (24) or (26) for $\beta_p \neq \beta_e \neq \beta_m$ or $\beta_p \neq \beta_e = \beta_m$, respectively. If $\beta_p = \beta_e \neq \beta_m$, then Equation (28) is replaced by the following mathematical condition:

$$\left(\varepsilon'_{22p}\right)_{r=R1} = \left(\varepsilon'_{22e}\right)_{r=R1} = -p_2 \rho_{1pe}, \tag{30}$$

where p_2 and ρ_{1pe} are given by Equations (27) and (28), (29), respectively. In this case, the integration constant C_p in $(\varepsilon'_{22p})_{r=R1}$ is a function of $(\varepsilon'_{22e})_{r=R1}$. The tangential strain $(\varepsilon'_{22e})_{r=R1}$ includes the integration constant C_e, which is determined by Equation (30).

Spherical Envelope

If $\beta_p \neq \beta_e \neq \beta_m$, then the mathematical boundary conditions are derived as:

$$\left(\sigma'_{1le}\right)_{r=R1} = -p_1, \tag{31}$$

$$\left(\sigma'_{1le}\right)_{r=R2} = -p_2, \tag{32}$$

where p_1 and p_2 are given by Equations (24) and (25), respectively. If $\beta_p \neq \beta_e = \beta_m$ or $\beta_p = \beta_e \neq \beta_m$, then the absolute value $|u'_{1e}|$ is required to represent a decreasing or increasing function of $r \in [R_1, R_2]$,with a maximum value for $r = R_1$. The mathematical boundary condition (31) or (32) is then considered, where p_1 or p_2 is given by Equation (26) or (27), respectively.

Cell Matrix

If $\beta_p \neq \beta_e \neq \beta_m$ or $\beta_p = \beta_e \neq \beta_m$, then the required mathematical boundary conditions are derived as:

$$\left(\sigma'_{1lm}\right)_{r=R2} = -p_2, \tag{33}$$

$$\left(u'_{1m}\right)_{r=rs} = 0, \tag{34}$$

where r_s and p_2 are given by Equations (2) and (26), respectively. If $\beta_p \neq \beta_e = \beta_m$, then Equation (33) is replaced by the following required mathematical boundary condition:

$$\left(\varepsilon'_{22m}\right)_{r=R2} = \left(\varepsilon'_{22e}\right)_{r=R2} = -p_1\rho_{1me}, \tag{35}$$

where ρ_{1me} is given by Equation (26), respectively. In this case, the integration constant C_m in $(\varepsilon'_{22m})_{r=R2}$ is a function of $(\varepsilon'_{22e})_{r=R2}$. The tangential strain $(\varepsilon'_{22e})_{r=R2}$ includes the integration constant C_e, which is determined by Equation (31). The model material system is conceptually divided into cubic cells (see Figure 1).

The surface *1234* of the neighbouring cubic cells *A* and *B* in Figure 4 is not a physical boundary. The functions $u'_{1A} = u'_{1A}(r, \varphi_A, \nu_A)$ and $u'_{1B} = u'_{1B}(r, \varphi B, \nu_B)$ are associated with cells A and B, respectively. These functions are connected at point *S*. This connection is assumed to be 'smooth'. Due to this assumption, the functions u'_{1A} and u'_{1B} of the variable r must not create a singular connection at point S. Consequently, S is not considered a singular point.

To satisfy this non-singularity assumption, the function $u'_{1m} = u'_{1m}(r, \varphi, \nu)$ of the variable $r \in \langle R_2, r_s \rangle$ must be extremal on the cell surface, i.e., at $r = r_s$. This extremum at $r = r_s$ represents a minimum due to the decreasing behavior of the function $|u'_{1m}|$. Based on Equation (3), the additional boundary condition for the cell matrix is derived as:

$$\left(\varepsilon'_{11m}\right)_{r=rs} = \left(\frac{\partial u'_{1m}}{\partial r}\right)_{r=rs} = 0. \tag{36}$$

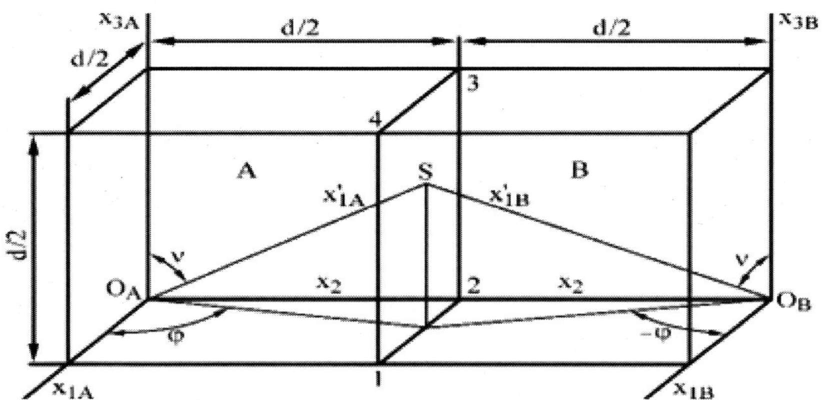

Figure 4. The arbitrary point *S* on the surface *1234* of the neighbouring cubic cells *A* and *B* with centres O_A and O_B, respectively, where $r_s = O_A S = O_B S$.

Superposition Method

The stresses induced around a specific spherical particle are influenced by those of the neighboring particles. As a result, the stresses in cell A are affected by those in cell B (see Figure 4). This influence from neighboring cells is determined by the radial boundary stress σ'_{1B}, which acts along the axis x'_{1A} at the point S, where σ'_{1B} represents the projections of $(\sigma'_{ijB})_{r=rs}$ (with $i,j = 1,2,3$) onto x'_{1A}. The angular position of point S relative to cells A and B is defined by φ, ν and - φ, ν, respectively. Consequently, we get:

$$\sigma'_{1B} = \sum_{i,j=1}^{3} \left(\sigma'_{ijB} \right)_{r=rs} \vec{e}'_{1A} \, \vec{e}'_{B} , \tag{37}$$

ē'$_{iQ}$ (where i = 1,2,3; Q = A, B) is a unit vector along the axis x_{iQ} (see Figure 2). Consequently, we get:

$$\vec{e}'_{i} = \sum_{j=1}^{3} e'_{ij} \vec{e}'_{j}, \quad i = 1,2,3 , \tag{38}$$

where ē$_i$ (i = 1,2,3) is a unit vector along the axis x_i, and $t_{ij} = \cos[\angle('_i, x_j)]$ (for i,j = 1,2,3) is derived as Brdicka et al. [5] and Ceniga [7]:

$$e'_{11} = \cos\varphi \sin\nu, \quad e'_{12} = \sin\varphi \sin\nu, \quad e'_{13} = e'_{33} = \cos\nu, \quad e'_{21} = -e'_{22} = -\sin\varphi,$$
$$e'_{23} = 0, \quad e'_{31} = -\cos\varphi\cos\nu, \quad e'_{32} = -\sin\varphi \cos\nu . \tag{39}$$

The radial boundary stress σ'_{1B} is derived as:

$$\sigma'_{1B} = \gamma_1 \left(\sigma'_{11m} \right)_{r=rs} + \gamma_2 \left(\sigma'_{22m} \right)_{r=rs} + \gamma_3 \left(\sigma'_{33m} \right)_{r=rs}$$
$$+ (\gamma_1 + \gamma_2) \left(\sigma'_{12m} \right)_{r=rs} + (\gamma_1 + \gamma_3) \left(\sigma'_{13m} \right)_{r=rs} , \tag{40}$$

where γ_i (for i = 1,2,3) is derived as:

$$\gamma_1 = 1 - 2\sin^2\varphi \sin^2\nu, \quad \gamma_2 = \sqrt{2} \sin\left(\frac{\pi}{4} - \varphi \right) \sin\varphi \sin\nu$$
$$\gamma_3 = \frac{1}{2} \sin\left(2\varphi - \frac{\pi}{2} \right) \sin 2\nu . \tag{41}$$

Regarding the superposition method, the total radial displacement u'_{1Tq}, the total stress σ'_{ijTq} and the total strain ε'_{ijTq} (for i,j = 1,2,3; q = p,e,m) are derived as:

$$u'_{1Tq1} = u'_{1q1} + u'_{1q2},\ \varepsilon'_{1Tq1} = \varepsilon'_{1q1} + \varepsilon'_{1q2},\ \sigma'_{1Tq1} = \sigma'_{1q1} + \sigma'_{1q2},$$

$$w_T = w_p + w_{pB} + w_e + w_{eB} + w_m + w_{mB},$$

$$W_T = W_p + W_{pB} + W_e + W_{eB} + W_m + W_{mB},$$

$$q_1 = p \Rightarrow q_2 = pB\ ,\ \ q_1 = e \Rightarrow q_2 = eB\ ,\ \ q_1 = m \Rightarrow q_2 = mB\ , \tag{42}$$

where σ'_{ijq1}, ε'_{ijq1} ,σ'_{ijq2}, and ε'_{ijq2} (i,j = 1,2,3), which induce w_{q1}, W_{q1} and w_{q2}, W_{q2}, are determined by the mathematical boundary conditions (28)-(36) and (43)-(46), respectively, where $\beta_p \neq \beta_e \neq \beta_m$, $\beta_p \neq \beta_e = \beta_m$, $\beta_p = \beta_e \neq \beta_m$ (Ceniga, [6]). Consequently, u'_{1mB}, σ'_{ijmB}, ε'_{ijmB} (for i,j = 1,2,3), w_{mB} and W_{mB}, which result from σ'_{1B}, are determined by the following mathematical boundary conditions:

$$\left(\sigma'_{11mB}\right)_{r=rs} = -\sigma'_{1B}, \tag{43}$$

$$\left(u'_{1mB}\right)_{r=rs} = 0. \tag{44}$$

Accordingly, u'_{1eB}, σ'_{ijeB}, ε'_{ijeB} (for i,j = 1,2,3), w_{eB}, W_{eB}, which result from σ'_{1B}, are determined by the following mathematical boundary condition:

$$\left(u'_{1eB}\right)_{r=R2} = \left(u'_{1mB}\right)_{r=R2}, \tag{45}$$

where the integration constant C_{eB} is a function of $(u'_{1mB})_{r=R2}$. Finally, u'_{1mB}, σ'_{ijmB}, ε'_{ijmB} (for i,j = 1,2,3), w_{mB}, and W_{mB}, which result from σ'_{1B}, are determined by the following mathematical boundary condition:

$$\left(u'_{1mB}\right)_{r=R1} = \left(u'_{1eB}\right)_{r=R1}, \tag{46}$$

where the integration constant C_{pB} is a function of $(u'_{1eB})_{r=R1}$, the absolute values $|u'_{1mB}|$ and $|u'_{1pB}|$ are decreasing and increasing functions of $r \in \langle R_2, r_s \rangle$ and $r \in \langle 0, R_1 \rangle$, respectively. The absolute value $|u'_{1eB}|$ is a decreasing or increasing function of $r \in \langle R_1, R_2 \rangle$. The conditions $(u'_{1pB})_{r=0} \neq \pm\infty, (\varepsilon'_{ijpB})_{r=0} \neq \pm\infty, (\sigma'_{ijpB})_{r=0} \neq \pm\infty$(for $i,j=1,2,3$) are required to be valid. The mathematical boundary condition (36) for ε'_{11mB}, along with Equation (44), is not applicable. Otherwise, we get $(\sigma'_{11mB})_{r=rs} \neq 0$, and the boundary condition (43) is not satisfied.

Mathematical Model of Thermal Stresses

Performing $\partial^2/\partial r^2$ in Equation (15), the differential Equation (15) is derived as:

$$r \frac{\partial^3 u_1'}{\partial r^3} + 4r^2 \frac{\partial^2 u_1'}{\partial r^2} + \frac{r}{s_{44}\left(c_1 + c_2\right)} \frac{\partial U_1'}{\partial r} = 0, \tag{47}$$

where s_{44}, c_1, c_2 and $U'_1 = U'_1(r, \varphi, \nu)$ are given by Equations (16), (11) and (17), respectively. By performing $\partial^2/\partial r^2$ in Equation (16), the differential Equation (16) is derived as:

$$r \frac{\partial^3 U_1'}{\partial r^3} + \left(2 - c_3\right) \frac{\partial^2 U_1'}{\partial r^2} = 0. \tag{48}$$

If U'_1 is assumed to be in the form $U'_1 = r^\lambda$, then the solution U'_1 of Equation (37) is derived as:

$$U_1' = \frac{\partial^2 U_1'}{\partial \varphi^2} + \frac{\partial^2 U_1'}{\partial \nu^2} = C_1 r + C_2 r^{c3} + C_3. \tag{49}$$

From Equations (16), (49), we obtain:

$$r \frac{\partial U_1'}{\partial r} = c_3 (C_1 r + C_2 r^{c3} + C_3). \tag{50}$$

Let Equation (50) be substituted to Equation (47), and then we get:

$$r^3 \frac{\partial^3 u_1'}{\partial r^3} + 4r^2 \frac{\partial^2 u_1'}{\partial r^2} = C_1 r^3 + C_2 r^{c3} + C_3 . \tag{51}$$

The mathematical solution of Equation (51), determined using the Wronskian method by Rektorys [9]:

$$u_1' = C_1 r\left(\frac{4}{3} - \ln r\right) + C_2 r^{c3} + C_3\left(\frac{1}{2} + \ln r\right), \tag{52}$$

where $c_3 < 0$ is given by Equation (11), and this mathematical solution is suitable for the spherical envelope and the cell matrix due to $(u'_{1p})_{r=0} = \pm\infty$ (see Equation (29)).With regard to Equations (3)-(10), (7)-(10), (18), we get:

$$\varepsilon_{11}' = C_1 \left(\frac{1}{3} - \ln r\right) + C_2 c_3 r^{c3-1} + \frac{C_3}{r},$$

$$\varepsilon_{22}' = \varepsilon_{33}' = C_1 \left(\frac{4}{3} - \ln r\right) + C_2 r^{c3-1} + \frac{C_3}{r}\left(\frac{1}{2} + \ln r\right),$$

$$\varepsilon_{12}' = s_{44}\sigma_{12}' = \left(\frac{4}{3} - \ln r\right)\frac{\partial C_1}{\partial \varphi} + r^{c3-1}\frac{\partial C_2}{\partial \varphi} + \frac{1}{r}\left(\frac{1}{2} + \ln r\right)\frac{\partial C_3}{\partial \varphi},$$

$$\varepsilon_{13}' = s_{44}\sigma_{13}' = \left(\frac{4}{3} - \ln r\right)\frac{\partial C_1}{\partial \nu} + r^{c3-1}\frac{\partial C_2}{\partial \nu} + \frac{1}{r}\left(\frac{1}{2} + \ln r\right)\frac{\partial C_3}{\partial \nu},$$

$$\sigma_{11}' = C_1 \left[\frac{c_1 - 7c_2}{3} - (c_1 - c_2)\ln r\right] + C_2 \left[(c_1 + c_2)c_3 - 2c_2\right] r^{c3-1}$$

$$+ \frac{C_3}{r}\left(c_1 - 2c_2 \ln r\right),$$

$$\sigma_{22}' = \sigma_{33}' = C_1 \left[\frac{4c_1 - c_2}{3} - (c_1 - c_2)\ln r\right] + C_2 \left(c_1 - c_2 c_3\right) r^{c3-1}$$

$$+ \frac{C_3}{r}\left(\frac{c_1 - 2c_2}{2} + c_1 \ln r\right),$$

$$w = C_1^2 \kappa_1 + C_2^2 \kappa_2 + C_3^2 \kappa_3 + C_1 C_2 \kappa_4 + C_1 C_3 \kappa_5 + C_2 C_3 \kappa_6$$

$$+\frac{\chi_1}{s_{44}}\left[\left(\frac{\partial C_1}{\partial \varphi}\right)^2+\left(\frac{\partial C_1}{\partial \nu}\right)^2\right]+\frac{\chi_2}{s_{44}}\left[\left(\frac{\partial C_2}{\partial \varphi}\right)^2+\left(\frac{\partial C_2}{\partial \nu}\right)^2\right]$$

$$+\frac{\chi_3}{s_{44}}\left[\left(\frac{\partial C_3}{\partial \varphi}\right)^2+\left(\frac{\partial C_3}{\partial \nu}\right)^2\right]+\frac{\chi_4}{s_{44}}\left[\frac{\partial C_1}{\partial \varphi}\frac{\partial C_2}{\partial \varphi}+\frac{\partial C_1}{\partial \nu}\frac{\partial C_2}{\partial \nu}\right]$$

$$+\frac{\chi_5}{s_{44}}\left[\frac{\partial C_1}{\partial \varphi}\frac{\partial C_3}{\partial \varphi}+\frac{\partial C_1}{\partial \nu}\frac{\partial C_3}{\partial \nu}\right]+\frac{\chi_6}{s_{44}}\left[\frac{\partial C_2}{\partial \varphi}\frac{\partial C_3}{\partial \varphi}+\frac{\partial C_2}{\partial \nu}\frac{\partial C_3}{\partial \nu}\right], \tag{53}$$

where κ_i, χ_i ($i = 1,...,6$) are derived as:

$$\kappa_1 = \frac{c_2 - c_1}{2}\ln^2 r + \frac{c_1 - c_2}{3}\ln r + \frac{17c_1 + c_2}{18},$$

$$\kappa_2 = \left[\frac{c_3^2(c_1 + c_2)}{2} + c_1(1 - 2c_3)\right] r^{2(c_3 - 1)},$$

$$\kappa_3 = \frac{1}{r^2}\left[c_1 \ln r(\ln r - 1) + \frac{c_2 - 2c_1}{4}\right],$$

$$\kappa_4 = \left\{c_3(c_1 - c_2)\ln r + \left[2c_1 + \frac{c_3(c_2 - 7c_1)}{3}\right]\right\} r^{c_3 - 1},$$

$$\kappa_5 = \left[(3c_1 - c_2)\ln r + \frac{4c_1 - c_2}{3}\right]\frac{1}{r},$$

$$\kappa_6 = [2c_1(1 - c_3)\ln r + (c_2 c_3 - c_1)]\, r^{c_3 - 2},$$

$$\chi_1 = \ln^2 r + \frac{8}{3}\ln r + \frac{16}{9}, \quad \chi_2 = r^{2(c_3 - 1)},$$

$$\chi_3 = \left(\ln^2 r + \ln r + \frac{1}{4}\right)\frac{1}{r^2}, \quad \chi_4 = 2\left(\frac{4}{3} - \ln r\right) r^{c_3 - 1},$$

$$\chi_5 = \left[\frac{1}{3}(4 + 5\ln r) - 2\ln^2 r\right]\frac{1}{r}, \quad \chi_6 = (2\ln r + 1)\, r^{c_3 - 2}. \tag{54}$$

The integrals Φ_{iq}, Ψ_{iq} ($i = 1,\ldots,6$) for the spherical envelope ($q = e$) and the cell matrix ($q = m$) are derived as:

$$\Phi_{iq} = \int_{r1}^{r2} \kappa_{iq} r^2 dr, \quad \Psi_{iq} = \int_{r1}^{r2} \chi_{iq} r^2 dr, \quad i, j = 1, \dots, 6,$$

$$q = e \Rightarrow r_1 = R_1,\ r_2 = R_2; \qquad q = m \Rightarrow r_1 = R_2,\ r_2 = r_S, \tag{55}$$

where r_s is given by Equation (2), and from Equations (54) and (55), we get:

$$\Phi_{1q} = \frac{c_{2q} - c_{1q}}{6}\left\{r_2^3\left[\left(\ln r_2 - \frac{1}{3}\right)^2 + \frac{1}{9}\right] - r_1^3\left[\left(\ln r_1 - \frac{1}{3}\right)^2 + \frac{1}{9}\right]\right\},$$

$$+\frac{c_{1q} - c_{2q}}{9}\left\{r_2^3\left(\ln r_2 - \frac{1}{3}\right) - r_1^3\left(\ln r_1 - \frac{1}{3}\right)\right\} + \frac{\left(17c_{1q} + c_{2q}\right)\left(r_2^3 - r_1^3\right)}{54},$$

$$\Phi_{2q} = \frac{1}{2c_{3q} + 1}\left[\frac{c_{3q}^2\left(c_{1q} + c_{2q}\right)}{2} + c_{1q}\left(1 - 2c_{3q}\right)\right]\left(r_2^{2c3q+1} - r_1^{2c3q+1}\right),$$

$$\Phi_{3q} = c_{1q}\left[r_2\left(\ln^2 r_2 - 2\ln r_2 + 2\right) - r_1\left(\ln^2 r_1 - 2\ln r_1 + 2\right)\right]$$

$$-c_{1q}\left[r_2\left(\ln r_2 - 1\right) - r_1\left(\ln r_1 - 1\right)\right] + \frac{\left(c_{2q} - 2c_{1q}\right)\left(r_2 - r_1\right)}{4},$$

$$\Phi_{4q} = \frac{c_{3q}\left(c_{1q} - c_{2q}\right)}{c_{3q} + 2}\left[r_2^{c3q+2}\left(\ln r_2 - \frac{1}{c_{3q} + 2}\right) - r_1^{c3q+2}\left(\ln r_1 - \frac{1}{c_{3q} + 2}\right)\right]$$

$$+\frac{1}{c_{3q} + 2}\left[2c_{1q} + \frac{c_{3q}\left(c_{2q} - 7c_{1q}\right)}{3}\right]\left(r_2^{c3q+2} - r_1^{c3q+2}\right),$$

$$\Phi_{5q} = \frac{3c_{1q} - c_{2q}}{2}\left[r_2^2\left(\ln r_2 - \frac{1}{2}\right) - r_1^2\left(\ln r_1 - \frac{1}{2}\right)\right] - \frac{\left(4c_{1q} - c_{2q}\right)\left(r_2^2 - r_1^2\right)}{6},$$

$$\Phi_{6q} = \frac{2c_{1q}\left(1 - c_{3q}\right)}{c_{3q} + 1}\left[r_2^{c3q+1}\left(\ln r_2 - \frac{1}{c_{3q} + 1}\right) - r_1^{c3q+1}\left(\ln r_1 - \frac{1}{c_{3q} + 1}\right)\right]$$

$$+\frac{\left(c_{2q}c_{3q} - c_{1q}\right)\left(r_2^{c3q+1} - r_1^{c3q+1}\right)}{c_{3q} + 1},$$

$$\Psi_{1q} = \frac{r_2^3}{3}\left[(\ln r_2 - 3)\left(\ln r_2 - \frac{1}{3}\right) + \frac{17}{9}\right] - \frac{r_1^3}{3}\left[(\ln r_1 - 3)\left(\ln r_1 - \frac{1}{3}\right) + \frac{17}{9}\right],$$

$$\Psi_{2q} = \frac{r_2^{2c3q+1} - r_1^{2c3q+1}}{2c_{3q} + 1},$$

$$\Psi_{3q} = r_2\left[\ln r_2 (\ln r_2 - 1) + \frac{5}{4}\right] - r_1\left[\ln r_1 (\ln r_1 - 1) + \frac{5}{4}\right],$$

$$\Psi_{4q} = \frac{2}{c_{3q} + 2}\left\{r_2^{c3q+2}\left[\frac{4c_{3q} + 11}{3(c_{3q} + 2)} - \ln r_2\right] - r_1^{c3q+2}\left[\frac{4c_{3q} + 11}{3(c_{3q} + 2)} - \ln r_1\right]\right\},$$

$$\Psi_{5q} = \frac{2\left(r_2^2 - r_1^2\right)}{3} + \frac{5}{6}\left[r_2^2\left(\ln r_2 - \frac{1}{2}\right) - r_1^2\left(\ln r_1 - \frac{1}{2}\right)\right]$$

$$- r_2^2\left(\ln^2 r_2 - \ln r_2 + \frac{1}{2}\right) + r_1^2\left(\ln^2 r_1 - \ln r_1 + \frac{1}{2}\right),$$

$$\Psi_{6q} = \frac{2}{c_{3q} + 1}\left[r_2^{c_{3q}+1}\left(\ln r_2 - \frac{1}{c_{3q} + 1} + \frac{1}{2}\right) - r_1^{c_{3q}+1}\left(\ln r_1 - \frac{1}{c_{3q} + 1} + \frac{1}{2}\right)\right],$$

$$q = e \Rightarrow r_1 = R_1,\ r_2 = R_2; \qquad q = m \Rightarrow r_1 = R_2,\ r_2 = r_S. \tag{56}$$

Condition $\beta_p \neq \beta_e \neq \beta_m$

Cell-Matrix: The integration constants C_{1m}, C_{2m}, C_{3m} (see Equation (52)) are determined by the mathematical boundary conditions (33), (34) or (33), (34), (36), which result in the following combinations: $C_{1m} \neq 0$, $C_{2m} \neq 0$, $C_{3m} = 0$; $C_{1m} \neq 0$, $C_{3m} \neq 0$, $C_{2m} = 0$; $C_{2m} \neq 0$, $C_{3m} \neq 0$, $C_{1m} = 0$; or $C_{1m} \neq 0$, $C_{2m} \neq 0$, $C_{3m} \neq 0$, respectively. Consequently, such a combination exhibits a minimum value of W_T (see Equation (42)). The subscript $q = m$, mB and the stress σ_q in this section are derived as:

$$q = m \Rightarrow \sigma_q = p_2; \qquad q = mB \Rightarrow \sigma_q = \sigma'_{1B}, \tag{57}$$

where p_2 and σ'_{1B} are given by Equations (25).

Condition $C_{1q} \neq 0,\ C_{2q} \neq 0,\ C_{3q} = 0$. From Equations (18), (32), (33), (34), (40), (43), (44), (52)-(56), we get:

$$\varepsilon'_{11q} = -\frac{\sigma_q}{\varsigma_q}\left[\frac{1}{3} - \ln r - c_{3m}\left(\frac{4}{3} - \ln r_S\right)\left(\frac{r}{r_S}\right)^{c3m-1}\right],$$

$$\varepsilon'_{22q} = \varepsilon'_{33q} = -\frac{\sigma_q}{\varsigma_q}\left[\frac{4}{3} - \ln r - \left(\frac{4}{3} - \ln r_S\right)\left(\frac{r}{r_S}\right)^{c3m-1}\right],$$

$$\varepsilon'_{12q} = s_{44m}\,\sigma'_{12q} = \left(\ln r - \frac{4}{3}\right)\frac{\partial}{\partial\varphi}\left(\frac{\sigma_q}{\varsigma_q}\right) + r^{c3m-1}\frac{\partial}{\partial\varphi}\left[\frac{\sigma_q}{\varsigma_q\, r_S^{c3m-1}}\left(\frac{4}{3} - \ln r_S\right)\right],$$

$$\varepsilon'_{13q} = s_{44m}\,\sigma'_{13q} = \left(\ln r - \frac{4}{3}\right)\frac{\partial}{\partial v}\left(\frac{\sigma_q}{\varsigma_q}\right) + r^{c3m-1}\frac{\partial}{\partial v}\left[\frac{\sigma_q}{\varsigma_q\, r_S^{c3m-1}}\left(\frac{4}{3} - \ln r_S\right)\right],$$

$$\sigma'_{11q} = \frac{\sigma_q}{\varsigma_q}\left\{\frac{7c_{2m} - c_{1m}}{3} + (c_{1m} - c_{2m})\ln r\right.$$

$$\left. + \left[(c_{1m} + c_{2m})c_{3m} - 2c_{2m}\right]\left(\frac{4}{3} - \ln r_s\right)\left(\frac{r}{r_s}\right)^{c3m-1}\right\},$$

$$\sigma'_{22q} = \sigma'_{33q} = \frac{\sigma_q}{\varsigma_q}\left[\frac{c_{2m} - 4c_{1m}}{3} + (c_{1m} - c_{2m})\ln r\right.$$

$$\left. + (c_{1m} - c_{2m}\ c_{3m})\left(\frac{4}{3} - \ln r_s\right)\left(\frac{r}{r_s}\right)^{c3m-1}\right],$$

$$w_q = \left(\frac{\sigma_q}{\varsigma_q}\right)^2\left[\kappa_{1m} + \kappa_{2m}\left(\frac{3\ln r_S - 4}{3r_S^{c3m-1}}\right)^2 + \frac{\kappa_{4m}(3\ln r_S - 4)}{3r_S^{c3m-1}}\right]$$

$$+ \frac{\chi_{1m}}{s_{44m}}\left[\left[\frac{\partial}{\partial\varphi}\left(\frac{\sigma_q}{\varsigma_q}\right)\right]^2 + \left[\frac{\partial}{\partial v}\left(\frac{\sigma_q}{\varsigma_q}\right)\right]^2\right]$$

$$+\frac{\chi_{2m}}{s_{44m}}\left(\left\{\frac{\partial}{\partial\varphi}\left[\frac{\sigma_q\left(3\ln r_S-4\right)}{3\varsigma_q\, r_S^{c3m-1}}\right]\right\}^2+\left\{\frac{\partial}{\partial\nu}\left[\frac{\sigma_q\left(3\ln r_S-4\right)}{3\varsigma_q\, r_S^{c3m-1}}\right]\right\}^2\right)$$

$$+\frac{\chi_{4m}}{s_{44m}}\left\{\frac{\partial}{\partial\varphi}\left(\frac{\sigma_q}{\varsigma_q}\right)\frac{\partial}{\partial\varphi}\left[\frac{\sigma_q\left(3\ln r_S-4\right)}{3\varsigma_q\, r_S^{c3m-1}}\right]+\frac{\partial}{\partial\nu}\left(\frac{\sigma_q}{\varsigma_q}\right)\frac{\partial}{\partial\nu}\left[\frac{\sigma_q\left(3\ln r_S-4\right)}{3\varsigma_q\, r_S^{c3m-1}}\right]\right\},$$

$$W_q=4\int_0^{\pi/2}\int_0^{\pi/2}\left(\frac{\sigma_q}{\varsigma_q}\right)^2\left[\Phi_{1m}+\Phi_{2m}\left(\frac{3\ln r_S-4}{3r_S^{c3m-1}}\right)^2+\frac{\Phi_{4m}\left(3\ln r_S-4\right)}{3\, r_S^{c3m-1}}\right]d\varphi\, d\nu$$

$$+\frac{4}{s_{44m}}\int_0^{\pi/2}\int_0^{\pi/2}\Psi_{1m}\left[\left[\frac{\partial}{\partial\varphi}\left(\frac{\sigma_q}{\varsigma_q}\right)\right]^2+\left[\frac{\partial}{\partial\nu}\left(\frac{\sigma_q}{\varsigma_q}\right)\right]^2\right]d\varphi\, d\nu$$

$$+\frac{4}{s_{44m}}\int_0^{\pi/2}\int_0^{\pi/2}\Psi_{2m}\left(\left\{\frac{\partial}{\partial\varphi}\left[\frac{\sigma_q\left(3\ln r_S-4\right)}{3\varsigma_q\, r_S^{c3m-1}}\right]\right\}^2\right.$$

$$\left.+\left\{\frac{\partial}{\partial\nu}\left[\frac{\sigma_q\left(3\ln r_S-4\right)}{3\varsigma_q\, r_S^{c3m-1}}\right]\right\}^2\right)d\varphi\, d\nu$$

$$+\frac{4}{s_{44m}}\int_0^{\pi/2}\int_0^{\pi/2}\Psi_{4m}\frac{\partial}{\partial\varphi}\left(\frac{\sigma_q}{\varsigma_q}\right)\frac{\partial}{\partial\varphi}\left[\frac{\sigma_q\left(3\ln r_S-4\right)}{3\varsigma_q\, r_S^{c3m-1}}\right]d\varphi\, d\nu$$

$$+\frac{4}{s_{44m}}\int_0^{\pi/2}\int_0^{\pi/2}\Psi_{4m}\frac{\partial}{\partial\nu}\left(\frac{\sigma_q}{\varsigma_q}\right)\frac{\partial}{\partial\nu}\left[\frac{\sigma_q\left(3\ln r_S-4\right)}{3\varsigma_q\, r_S^{c3m-1}}\right]d\varphi\, d\nu,$$

$$\sigma_{1B}'=-\frac{\rho_{1B}p_2}{\varsigma_m}-\frac{1}{s_{44m}}\left[\rho_{1B}^{(\varphi)}\frac{\partial}{\partial\varphi}\left(\frac{p_2}{\varsigma_m}\right)+\rho_{1B}^{(\nu)}\frac{\partial}{\partial\nu}\left(\frac{p_2}{\varsigma_m}\right)\right]$$

$$+\frac{\rho_{2B}p_2}{\varsigma_m r_S^{c3m-1}}\left(\frac{4}{3}-\ln r_S\right)+\frac{\rho_{2B}^{(\varphi)}}{s_{44m}}\frac{\partial}{\partial\varphi}\left[\frac{p_2}{\varsigma_m r_S^{c3m-1}}\left(\frac{4}{3}-\ln r_S\right)\right]$$

$$+\frac{\rho_{2B}^{(v)}}{s_{44m}}\frac{\partial}{\partial v}\left[\frac{p_2}{\varsigma_m r_S^{c_{3m}-1}}\left(\frac{4}{3}-\ln r_S\right)\right], \tag{58}$$

where r_S s_{44m}; c_{im} $(i = 1,2,3)$, κ_{jm}, χ_{jm}; Φ_{jm}, Ψ_{jm} $(j = 1,2,4)$ are given by Equations (2), (16), (11), (54), (56), respectively. The coefficients ς_q, ς_{iq}, ρ_{iB}, $\rho_{iB}^{(\tau)}$ $(i = 1,2;\ \tau = \varphi, v)$ are derived as:

$$\varsigma_q = \varsigma_{2q} - \varsigma_{1q}\left(\frac{4}{3}-\ln r_S\right), \varsigma_{1m} = \varsigma_{1mB}\left(\frac{R_2}{r_S}\right)^{c_{3m}-1},$$

$$\varsigma_{2m} = \frac{c_{1m}-7c_{2m}}{3} - (c_{1m}-c_{2m})\ln R_2,$$

$$\varsigma_{1mB} = (c_{1m}+c_{2m})c_{3m} - 2c_{2m},\ \varsigma_{2mB} = \frac{c_{1m}-7c_{2m}}{3} - (c_{1m}-c_{2m})\ln r_S,$$

$$\rho_{1B} = \frac{1}{3}\{\gamma_1(c_{1m}+c_{2m})c_{3m} + (\gamma_2+\gamma_3)(4c_{1m}-c_{2m})\}$$

$$+(c_{1m}+c_{2m})(\gamma_1+\gamma_2+\gamma_3)\ln r_S,$$

$$\rho_{2B} = \{\gamma_1[(c_{1m}+c_{2m})c_{3m}-2c_{2m}] + (\gamma_2+\gamma_3)(c_{1m}-c_{2m}c_{3m})\}r_S^{c_{3m}-1},$$

$$\rho_{1B}^{(\varphi)} = (\gamma_1+\gamma_2)\left(\frac{4}{3}-\ln r_S\right),\ \rho_{1B}^{(v)} = (\gamma_1+\gamma_3)\left(\frac{4}{3}-\ln r_S\right),$$

$$\rho_{2B}^{(\varphi)} = (\gamma_1+\gamma_2)r_S^{c_{3m}-1},\ \rho_{21B}^{(v)} = (\gamma_1+\gamma_3)r_S^{c_{3m}-1}. \tag{59}$$

where γ_i $(i = 1,2,3)$ is given by Equation (41). From $(\varepsilon'_{22q})_{r=R2} = -\ \sigma_q\ \rho_q$ and Equation (58), the coefficient q is derived as:

$$\rho_q = \frac{1}{\varsigma_q}\left[\frac{4}{3}-\ln R_2 - \left(\frac{4}{3}-\ln r_S\right)\left(\frac{R_2}{r_S}\right)^{c_{3m}-1}\right]. \tag{60}$$

Condition $C_{1q} \neq 0,\ C_{3q} \neq 0,\ C_{2q} = 0$. Similarly, from Equations (18), (32), (33), (34), (40), (43), (44), (52)-(56), we get:

$$\varepsilon_{11q}^{'} = -\frac{\sigma_q}{\varsigma_q}\left[\left(\frac{1}{3} - \ln r\right)\left(\frac{1}{2} + \ln r_S\right) - \frac{r_S}{r}\left(\frac{4}{3} - \ln r_S\right)\right],$$

$$\varepsilon_{22q}^{'} = \varepsilon_{33q}^{'} = -\frac{\sigma_q}{\varsigma_q}\left[\left(\frac{4}{3} - \ln r\right)\left(\frac{1}{2} + \ln r_S\right) - \frac{r_S}{r}\left(\frac{1}{2} + \ln r\right)\left(\frac{4}{3} - \ln r_S\right)\right],$$

$$\varepsilon_{12q}^{'} = s_{44m}\,\sigma_{12q}^{'} = \left(\ln r - \frac{4}{3}\right)\frac{\partial}{\partial \varphi}\left[\frac{\sigma_q}{\varsigma_q}\left(\frac{1}{2} + \ln r_S\right)\right]$$

$$+\frac{1}{r}\left(\frac{1}{2} + \ln r\right)\frac{\partial}{\partial \varphi}\left[\frac{\sigma_q\, r_S}{\varsigma_q}\left(\ln r_S - \frac{4}{3}\right)\right],$$

$$\varepsilon_{13q}^{'} = s_{44m}\,\sigma_{13q}^{'} = \left(\ln r - \frac{4}{3}\right)\frac{\partial}{\partial \nu}\left[\frac{\sigma_q}{\varsigma_q}\left(\frac{1}{2} + \ln r_S\right)\right]$$

$$+\frac{1}{r}\left(\frac{1}{2} + \ln r\right)\frac{\partial}{\partial \nu}\left[\frac{\sigma_q\, r_S}{\varsigma_q}\left(\ln r_S - \frac{4}{3}\right)\right],$$

$$\sigma_{11q}^{'} = -\frac{\sigma_q}{\varsigma_q}\left\{\left[\frac{7c_{2m} - c_{1m}}{3} + \left(c_{1m} - c_{2m}\right)\ln r\right]\left(\frac{1}{2} + \ln r_S\right)\right.$$

$$\left. + \frac{r_S}{r}\left(c_{1m} - 2c_{2m}\ln r\right)\left(\ln r_s - \frac{4}{3}\right)\right\},$$

$$\sigma_{22q}^{'} = \sigma_{33q}^{'} = -\frac{\sigma_q}{\varsigma_q}\left\{\left[\frac{c_{2m} - 4c_{1m}}{3} + \left(c_{1m} - c_{2m}\right)\ln r\right]\left(\frac{1}{2} + \ln r_S\right)\right.$$

$$\left. + \frac{r_s}{r}\left(\frac{c_{1m} - 2c_{2m}}{2} + c_{1m}\ln r\right)\left(\ln r_s - \frac{4}{3}\right)\right\},$$

$$w_q = \left(\frac{\sigma_q}{\varsigma_q}\right)^2\left\{\kappa_{1m}\left(\frac{1}{2} + \ln r_S\right)\right.$$

$$\left. + r_S\left(\frac{4}{3} - \ln r_S\right)\left[\kappa_{3m}\, r_S\left(\frac{4}{3} - \ln r_S\right) + \kappa_{5m}\left(\frac{1}{2} + \ln r_S\right)\right]\right\}$$

$$+\frac{\chi_{1m}}{s_{44m}}\left(\left\{\frac{\partial}{\partial \varphi}\left[\frac{\sigma_q}{\varsigma_q}\left(\frac{1}{2} + \ln r_S\right)\right]\right\}^2 + \left\{\frac{\partial}{\partial \nu}\left[\frac{\sigma_q}{\varsigma_q}\left(\frac{1}{2} + \ln r_S\right)\right]\right\}^2\right)$$

$$+\frac{\chi_{3m}}{s_{44m}}\left(\left\{\frac{\partial}{\partial\varphi}\left[\frac{\sigma_q r_S}{\varsigma_q}\left(\frac{4}{3}-\ln r_S\right)\right]\right\}^2+\left\{\frac{\partial}{\partial v}\left[\frac{\sigma_q r_S}{\varsigma_q}\left(\frac{4}{3}-\ln r_S\right)\right]\right\}^2\right)$$

$$+\frac{\chi_{5m}}{s_{44m}}\left\{\frac{\partial}{\partial\varphi}\left[\frac{\sigma_q}{\varsigma_q}\left(\frac{1}{2}+\ln r_S\right)\right]\frac{\partial}{\partial\varphi}\left[\frac{\sigma_q r_S}{\varsigma_q}\left(\frac{4}{3}-\ln r_S\right)\right]\right.$$

$$\left.+\frac{\partial}{\partial v}\left[\frac{\sigma_q}{\varsigma_q}\left(\frac{1}{2}+\ln r_S\right)\right]\frac{\partial}{\partial v}\left[\frac{\sigma_q r_S}{\varsigma_q}\left(\frac{4}{3}-\ln r_S\right)\right]\right\},$$

$$W_q=4\int_0^{\pi/2}\int_0^{\pi/2}\left(\frac{\sigma_q}{\varsigma_q}\right)^2\left\{\Phi_{1m}\left(\frac{1}{2}+\ln r_S\right)+\Phi_{3m}\left[r_S\left(\frac{4}{3}-\ln r_S\right)\right]^2\right.$$

$$\left.+\Phi_{5m}\,r_S\left(\frac{1}{2}+\ln r_S\right)\left(\frac{4}{3}-\ln r_S\right)\right\}d\varphi\,dv$$

$$+\frac{4}{s_{44m}}\int_0^{\pi/2}\int_0^{\pi/2}\Psi_{1m}\left(\left\{\frac{\partial}{\partial\varphi}\left[\frac{\sigma_q}{\varsigma_q}\left(\frac{1}{2}+\ln r_S\right)\right]\right\}^2\right.$$

$$\left.+\left\{\frac{\partial}{\partial v}\left[\frac{\sigma_q}{\varsigma_q}\left(\frac{1}{2}+\ln r_S\right)\right]\right\}^2\right)d\varphi\,dv$$

$$+\frac{4}{s_{44m}}\int_0^{\pi/2}\int_0^{\pi/2}\Psi_{3m}\left(\left\{\frac{\partial}{\partial\varphi}\left[\frac{\sigma_q r_S}{\varsigma_q}\left(\frac{4}{3}-\ln r_S\right)\right]\right\}^2\right.$$

$$\left.+\left\{\frac{\partial}{\partial v}\left[\frac{\sigma_q r_S}{\varsigma_q}\left(\frac{4}{3}-\ln r_S\right)\right]\right\}^2\right)d\varphi\,dv$$

$$+\frac{4}{s_{44m}}\int_0^{\pi/2}\int_0^{\pi/2}\Psi_{5m}\frac{\partial}{\partial\varphi}\left[\frac{\sigma_q}{\varsigma_q}\left(\frac{1}{2}+\ln r_S\right)\right]\frac{\partial}{\partial\varphi}\left[\frac{\sigma_q r_S}{\varsigma_q}\left(\frac{4}{3}-\ln r_S\right)\right]d\varphi\,dv$$

$$+\frac{4}{s_{44m}}\int_0^{\pi/2}\int_0^{\pi/2}\Psi_{5m}\frac{\partial}{\partial v}\left[\frac{\sigma_q}{\varsigma_q}\left(\frac{1}{2}+\ln r_S\right)\right]\frac{\partial}{\partial v}\left[\frac{\sigma_q r_S}{\varsigma_q}\left(\frac{4}{3}-\ln r_S\right)\right]d\varphi\,dv,$$

$$\sigma_{1B}^{'} = -\frac{\rho_{1B} p_2}{\varsigma_m}\left(\frac{1}{2} + \ln r_S\right) - \frac{\rho_{3B} p_2 r_S}{\varsigma_m}\left(\frac{4}{3} - \ln r_S\right)$$

$$-\frac{1}{s_{44m}}\left\{\rho_{1B}^{(\varphi)} \frac{\partial}{\partial \varphi}\left[\frac{p_2}{\varsigma_m}\left(\frac{1}{2} + \ln r_S\right)\right] + \rho_{3B}^{(\nu)} \frac{\partial}{\partial \nu}\left[\frac{p_2}{\varsigma_m}\left(\frac{1}{2} + \ln r_S\right)\right]\right\}$$

$$-\frac{1}{s_{44m}}\left\{\rho_{3B}^{(\varphi)} \frac{\partial}{\partial \varphi}\left[\frac{p_2 r_S}{\varsigma_m}\left(\frac{4}{3} - \ln r_S\right)\right] + \rho_{3B}^{(\nu)} \frac{\partial}{\partial \nu}\left[\frac{p_2 r_S}{\varsigma_m}\left(\frac{4}{3} - \ln r_S\right)\right]\right\},$$

$$\rho_q = \frac{1}{\varsigma_q}\left[\left(\frac{4}{3} - \ln R_2\right)\left(\frac{1}{2} + \ln r_S\right) - \frac{r_S}{R_2}\left(\frac{4}{3} - \ln r_S\right)\left(\frac{1}{2} + \ln R_2\right)\right],$$

$$\varsigma_q = \frac{\varsigma_{2q}}{r_S}\left(\frac{1}{2} + \ln r_S\right) - \varsigma_{1q}\left(\frac{4}{3} - \ln r_S\right),$$

$$\varsigma_{1m} = \frac{r_S}{R_2}\left(c_{1m} - 2c_{2m} \ln R_2\right), \quad \varsigma_{2m} = r_S\left[\frac{c_{1m} - 7c_{2m}}{3} - \left(c_{1m} - c_{2m}\right)\ln R_2\right],$$

$$\varsigma_{1mB} = c_{1m} - 2c_{2m} \ln r_S, \quad \varsigma_{2mB} = r_S\left[\frac{c_{1m} - 7c_{2m}}{3} - \left(c_{1m} - c_{2m}\right)\ln r_S\right],$$

$$\rho_{3B} = \frac{1}{r_S}\left\{c_{1m}\gamma_1 + \frac{\left(c_{1m} - 2c_{2m}\right)\left(\gamma_2 + \gamma_3\right)}{2} + \left[c_{1m}\left(\gamma_2 + \gamma_3\right) - 2c_{2m}\gamma_1\right]\ln r_S\right\},$$

$$\rho_{3B}^{(\varphi)} = \frac{\left(\gamma_1 + \gamma_2\right)}{r_S}\left(\frac{1}{2} + \ln r_S\right), \quad \rho_{3B}^{(\nu)} = \frac{\left(\gamma_1 + \gamma_3\right)}{r_S}\left(\frac{1}{2} + \ln r_S\right), \tag{61}$$

where ρ_{1B}, $\rho_{1B}^{(\tau)}$ ($\tau = \varphi, \nu$) are given by Equation (59).

Condition $C_{2q} \neq 0$, $C_{3q} \neq 0$, $C_{1q} = 0$. Similarly, from Equations (18), (32), (33), (34), (40), (43), (44), (52)-(56), we get:

$$\varepsilon_{11q}^{'} = -\frac{\sigma_q}{\varsigma_q}\left[c_{3m}\left(\frac{1}{2} + \ln r_S\right)r^{c_{3m}-1} - \frac{r_S^{c_{3m}}}{r}\right],$$

$$\varepsilon_{22q}^{'} = \varepsilon_{33q}^{'} = -\frac{\sigma_q}{\varsigma_q}\left[\left(\frac{1}{2} + \ln r_S\right)r^{c_{3m}-1} - \frac{r_S^{c_{3m}}}{r}\left(\frac{1}{2} + \ln r\right)\right],$$

$$\varepsilon_{12q}^{'} = s_{44m}\sigma_{12q}^{'} = \frac{1}{r}\left(\frac{1}{2} + \ln r\right)\frac{\partial}{\partial \varphi}\left(\frac{\sigma_q r_S^{c_{3m}}}{\varsigma_q}\right) - r^{c_{3m}-1}\frac{\partial}{\partial \varphi}\left[\frac{\sigma_q}{\varsigma_q}\left(\frac{1}{2} + \ln r_S\right)\right],$$

$$\varepsilon'_{13q} = s_{44m}\sigma'_{13q} = \frac{1}{r}\left(\frac{1}{2}+\ln r\right)\frac{\partial}{\partial v}\left(\frac{\sigma_q r_S^{c3m}}{\varsigma_q}\right) - r^{c3m-1}\frac{\partial}{\partial v}\left[\frac{\sigma_q}{\varsigma_q}\left(\frac{1}{2}+\ln r_S\right)\right],$$

$$\sigma'_{11q} = -\frac{\sigma_q r^{c3m-1}}{\varsigma_q}\left\{\left[c_{3m}\left(c_{1m}+c_{2m}\right)-2c_{2m}\right]\left(\frac{1}{2}+\ln r_S\right)\right.$$

$$\left.-\frac{r_S}{r}\left(c_{1m}-2c_{2m}\ln r\right)\right\}$$

$$\sigma'_{22q} = \sigma'_{33q} = -\frac{\sigma_q r^{c3m-1}}{\varsigma_q}\left\{\left(c_{1m}-c_{2m}c_{3m}\right)\left(\frac{1}{2}+\ln r_S\right)\right.$$

$$\left.-\frac{r_s}{r}\left(\frac{c_{1m}-2c_{2m}}{2}+c_{1m}\ln r\right)\right\},$$

$$w_q = \left(\frac{\sigma_q}{\varsigma_q}\right)^2\left[\kappa_{2m}\left(\frac{1}{2}+\ln r_S\right)^2 + \kappa_{3m} r_S^{2c3m} + \kappa_{6m} r_S^{c3m}\left(\frac{1}{2}+\ln r_S\right)\right]$$

$$+\frac{\chi_{2m}}{s_{44m}}\left(\left\{\frac{\partial}{\partial\varphi}\left[\frac{\sigma_q}{\varsigma_q}\left(\frac{1}{2}+\ln r_S\right)\right]\right\}^2 + \left\{\frac{\partial}{\partial v}\left[\frac{\sigma_q}{\varsigma_q}\left(\frac{1}{2}+\ln r_S\right)\right]\right\}^2\right)$$

$$+\frac{\chi_{3m}}{s_{44m}}\left\{\left[\frac{\partial}{\partial\varphi}\left(\frac{\sigma_q r_S^{c3m}}{\varsigma_q}\right)\right]^2 + \left[\frac{\partial}{\partial v}\left(\frac{\sigma_q r_S^{c3m}}{\varsigma_q}\right)\right]^2\right\}$$

$$-\frac{\chi_{6m}}{s_{44m}}\left\{\frac{\partial}{\partial\varphi}\left[\frac{\sigma_q}{\varsigma_q}\left(\frac{1}{2}+\ln r_S\right)\right]\frac{\partial}{\partial\varphi}\left(\frac{\sigma_q r_S^{c3m}}{\varsigma_q}\right)\right.$$

$$\left.+\frac{\partial}{\partial v}\left[\frac{\sigma_q}{\varsigma_q}\left(\frac{1}{2}+\ln r_S\right)\right]\frac{\partial}{\partial v}\left(\frac{\sigma_q r_S^{c3m}}{\varsigma_q}\right)\right\},$$

$$W_q = 4\int_0^{\pi/2}\int_0^{\pi/2}\left(\frac{\sigma_q}{\varsigma_q}\right)^2\left[\Phi_{2m}\left(\frac{1}{2}+\ln r_S\right)^2 + \Phi_{3m} r_S^{2c3m}\right.$$

$$\left.+\Phi_{6m} r_S^{c3m}\left(\frac{1}{2}+\ln r_S\right)\right] d\varphi\, dv$$

$$+\frac{4}{s_{44m}}\int_{0}^{\pi/2}\int_{0}^{\pi/2}\Psi_{2m}\left(\left\{\frac{\partial}{\partial\varphi}\left[\frac{\sigma_q}{\varsigma_q}\left(\frac{1}{2}+\ln r_S\right)\right]\right\}^2\right.$$

$$\left.+\left\{\frac{\partial}{\partial v}\left[\frac{\sigma_q}{\varsigma_q}\left(\frac{1}{2}+\ln r_S\right)\right]\right\}^2\right)d\varphi\, dv$$

$$+\frac{4}{s_{44m}}\int_{0}^{\pi/2}\int_{0}^{\pi/2}\Psi_{3m}\left\{\left[\frac{\partial}{\partial\varphi}\left(\frac{\sigma_q r_S^{c3m}}{\varsigma_q}\right)\right]^2+\left[\frac{\partial}{\partial v}\left(\frac{\sigma_q r_S^{c3m}}{\varsigma_q}\right)\right]^2\right\}d\varphi\, dv$$

$$+\frac{4}{s_{44m}}\int_{0}^{\pi/2}\int_{0}^{\pi/2}\Psi_{6m}\left\{\frac{\partial}{\partial\varphi}\left[\frac{\sigma_q}{\varsigma_q}\left(\frac{1}{2}+\ln r_S\right)\right]\frac{\partial}{\partial\varphi}\left(\frac{\sigma_q r_S^{c3m}}{\varsigma_q}\right)\right.$$

$$\left.+\frac{\partial}{\partial v}\left[\frac{\sigma_q}{\varsigma_q}\left(\frac{1}{2}+\ln r_S\right)\right]\frac{\partial}{\partial v}\left(\frac{\sigma_q r_S^{c3m}}{\varsigma_q}\right)\right\}d\varphi\, dv,$$

$$\sigma'_{1B}=-\frac{\rho_{1B}p_2}{\varsigma_m}\left(\frac{1}{2}+\ln r_S\right)+\frac{\rho_{3B}p_2 r_S^{c3m}}{\varsigma_m}$$

$$-\frac{1}{s_{44m}}\left\{\rho_{1B}^{(\varphi)}\frac{\partial}{\partial\varphi}\left[\frac{p_2}{\varsigma_m}\left(\frac{1}{2}+\ln r_S\right)\right]+\rho_{3B}^{(v)}\frac{\partial}{\partial v}\left[\frac{p_2}{\varsigma_m}\left(\frac{1}{2}+\ln r_S\right)\right]\right\}$$

$$+\frac{1}{s_{44m}}\left[\rho_{3B}^{(\varphi)}\frac{\partial}{\partial\varphi}\left(\frac{p_2 r_S^{c3m}}{\varsigma_m}\right)+\rho_{3B}^{(v)}\frac{\partial}{\partial v}\left(\frac{p_2 r_S^{c3m}}{\varsigma_m}\right)\right],$$

$$\rho_q=\frac{1}{\varsigma_q}\left[\left(\frac{1}{2}+\ln r_S\right)R_2^{c3m-1}-\frac{r_S^{c3m}}{R_2}\left(\frac{1}{2}+\ln R_2\right)\right],$$

$$\varsigma_q=\frac{\varsigma_{2q}}{r_S}\left(\frac{1}{2}+\ln r_S\right)-\varsigma_{1q}r_S^{c3m-1},$$

$$\varsigma_{1m}=\frac{r_S}{R_2}\left(c_{1m}-2c_{2m}\ln R_2\right),\quad \varsigma_{2m}=\left[c_{3m}\left(c_{1m}+c_{2m}\right)-2c_{2m}\right]r_S\, R_2^{c3m-1},$$

$$\varsigma_{1mB}=c_{1m}-2c_{2m}\ln r_S,\quad \varsigma_{2mB}=\left[c_{3m}\left(c_{1m}+c_{2m}\right)-2c_{2m}\right]r_S^{c3m-1} \tag{62}$$

Conditions $C_{1m} \neq 0$, $C_{2m} \neq 0$, $C_{3m} \neq 0$. Finally, from Equations (18), (32), (33), (34), (36), (52)-(56), we get:

$$\varepsilon'_{11m} = -\frac{p_2}{\varsigma_m}\left[\varsigma_{1m}\left(\frac{1}{3} - \ln r\right) - \varsigma_{2m} c_{3m} r^{c_{3m}-1} + \frac{\varsigma_{3m}}{r}\right],$$

$$\varepsilon'_{22m} = \varepsilon'_{33m} = \frac{p_2}{\varsigma_m}\left[\varsigma_{1m}\left(\frac{4}{3} - \ln r\right) + \varsigma_{2m} r^{c_{3m}-1} + \frac{\varsigma_{3m}}{r}\left(\frac{1}{2} + \ln r\right)\right],$$

$$\varepsilon'_{12m} = s_{44m}\,\sigma'_{12m} = -\left[\left(\frac{4}{3} - \ln r\right)\frac{\partial}{\partial\varphi}\left(\frac{p_2\varsigma_{1m}}{\varsigma_m}\right) + r^{c_{3m}-1}\frac{\partial}{\partial\varphi}\left(\frac{p_2\varsigma_{2m}}{\varsigma_m}\right)\right.$$

$$\left. + \frac{1}{r}\left(\frac{1}{2} + \ln r\right)\frac{\partial}{\partial\varphi}\left(\frac{p_2\varsigma_{3m}}{\varsigma_m}\right)\right],$$

$$\varepsilon'_{13m} = s_{44m}\,\sigma'_{13m} = -\left[\left(\frac{4}{3} - \ln r\right)\frac{\partial}{\partial v}\left(\frac{p_2\varsigma_{1m}}{\varsigma_m}\right) + r^{c_{3m}-1}\frac{\partial}{\partial v}\left(\frac{p_2\varsigma_{2m}}{\varsigma_m}\right)\right.$$

$$\left. + \frac{1}{r}\left(\frac{1}{2} + \ln r\right)\frac{\partial}{\partial v}\left(\frac{p_2\varsigma_{3m}}{\varsigma_m}\right)\right],$$

$$\sigma'_{11m} = -\frac{p_2}{\varsigma_m}\left\{\varsigma_{1m}\left[\frac{c_{1m} - 7c_{2m}}{3} - (c_{1m} - c_{2m})\ln r\right]\right.$$

$$\left. + \varsigma_{2m}\left[(c_{1m} + c_{2m})c_{3m} - 2c_{2m}\right]r^{c_{3m}-1} + \frac{\varsigma_{3m}(c_{1m} - 2c_{2m}\ln r)}{r}\right\},$$

$$\sigma'_{22m} = \sigma'_{33m} = -\frac{p_2}{\varsigma_m}\left\{\varsigma_{1m}\left[\frac{4c_{1m} - c_{2m}}{3} - (c_{1m} - c_{2m})\ln r\right]\right.$$

$$\left. + \varsigma_{2m}(c_{1m} - c_{2m}c_{3m})r^{c_{3m}-1} + \varsigma_{3m}\left(\frac{c_{1m} - 2c_{2m}}{2} + c_{1m}\ln r\right)\right\}$$

$$w_m = \left(\frac{p_2}{\varsigma_m}\right)^2\left(\kappa_{1m}\varsigma_{1m}^2 + \kappa_{2m}\varsigma_{2m}^2 + \kappa_{3m}\varsigma_{3m}^2\right.$$

$$\left. + \kappa_{4m}\varsigma_{1m}\varsigma_{2m} + \kappa_{5m}\varsigma_{1m}\varsigma_{3m} + \kappa_{6m}\varsigma_{2m}\varsigma_{3m}\right)$$

$$+ \frac{\chi_{1m}}{s_{44m}}\left\{\left[\frac{\partial}{\partial\varphi}\left(\frac{p_2\varsigma_{1m}}{\varsigma_m}\right)\right]^2 + \left[\frac{\partial}{\partial v}\left(\frac{p_2\varsigma_{1m}}{\varsigma_m}\right)\right]^2\right\}$$

$$+\frac{\chi_{2m}}{s_{44m}}\left\{\left[\frac{\partial}{\partial\varphi}\left(\frac{p_2\varsigma_{2m}}{\varsigma_m}\right)\right]^2+\left[\frac{\partial}{\partial v}\left(\frac{p_2\varsigma_{2m}}{\varsigma_m}\right)\right]^2\right\}$$

$$+\frac{\chi_{3m}}{s_{44m}}\left\{\left[\frac{\partial}{\partial\varphi}\left(\frac{p_2\varsigma_{3m}}{\varsigma_m}\right)\right]^2+\left[\frac{\partial}{\partial v}\left(\frac{p_2\varsigma_{3m}}{\varsigma_m}\right)\right]^2\right\}$$

$$+\frac{\chi_{4m}}{s_{44m}}\left[\frac{\partial}{\partial\varphi}\left(\frac{p_2\varsigma_{1m}}{\varsigma_m}\right)\frac{\partial}{\partial\varphi}\left(\frac{p_2\varsigma_{2m}}{\varsigma_m}\right)+\frac{\partial}{\partial v}\left(\frac{p_2\varsigma_{1m}}{\varsigma_m}\right)\frac{\partial}{\partial v}\left(\frac{p_2\varsigma_{2m}}{\varsigma_m}\right)\right]$$

$$+\frac{\chi_{5m}}{s_{44m}}\left[\frac{\partial}{\partial\varphi}\left(\frac{p_2\varsigma_{1m}}{\varsigma_m}\right)\frac{\partial}{\partial\varphi}\left(\frac{p_2\varsigma_{3m}}{\varsigma_m}\right)+\frac{\partial}{\partial v}\left(\frac{p_2\varsigma_{1m}}{\varsigma_m}\right)\frac{\partial}{\partial v}\left(\frac{p_2\varsigma_{3m}}{\varsigma_m}\right)\right]$$

$$+\frac{\chi_{6m}}{s_{44m}}\left[\frac{\partial}{\partial\varphi}\left(\frac{p_2\varsigma_{2m}}{\varsigma_m}\right)\frac{\partial}{\partial\varphi}\left(\frac{p_2\varsigma_{3m}}{\varsigma_m}\right)+\frac{\partial}{\partial v}\left(\frac{p_2\varsigma_{2m}}{\varsigma_m}\right)\frac{\partial}{\partial v}\left(\frac{p_2\varsigma_{3m}}{\varsigma_m}\right)\right],$$

$$W_m=4\int_0^{\pi/2}\int_0^{\pi/2}\left(\frac{p_2}{\varsigma_m}\right)^2\left(\Phi_{1m}\varsigma_{1m}^2+\Phi_{2m}\varsigma_{2m}^2+\Phi_{3m}\varsigma_{3m}^2\right.$$

$$\left.+\Phi_{4m}\varsigma_{1m}\varsigma_{2m}+\Phi_{5m}\varsigma_{1m}\varsigma_{3m}+\Phi_{6m}\varsigma_{2m}\varsigma_{3m}\right)d\varphi\,dv$$

$$+\frac{4}{s_{44m}}\int_0^{\pi/2}\int_0^{\pi/2}\frac{\Psi_{1m}}{s_{44m}}\left\{\left[\frac{\partial}{\partial\varphi}\left(\frac{p_2\varsigma_{1m}}{\varsigma_m}\right)\right]^2+\left[\frac{\partial}{\partial v}\left(\frac{p_2\varsigma_{1m}}{\varsigma_m}\right)\right]^2\right\}d\varphi\,dv$$

$$+\frac{4}{s_{44m}}\int_0^{\pi/2}\int_0^{\pi/2}\frac{\Psi_{2m}}{s_{44m}}\left\{\left[\frac{\partial}{\partial\varphi}\left(\frac{p_2\varsigma_{2m}}{\varsigma_m}\right)\right]^2+\left[\frac{\partial}{\partial v}\left(\frac{p_2\varsigma_{2m}}{\varsigma_m}\right)\right]^2\right\}d\varphi\,dv$$

$$+\frac{4}{s_{44m}}\int_0^{\pi/2}\int_0^{\pi/2}\frac{\Psi_{3m}}{s_{44m}}\left\{\left[\frac{\partial}{\partial\varphi}\left(\frac{p_2\varsigma_{3m}}{\varsigma_m}\right)\right]^2+\left[\frac{\partial}{\partial v}\left(\frac{p_2\varsigma_{3m}}{\varsigma_m}\right)\right]^2\right\}d\varphi\,dv$$

$$+\frac{4}{s_{44m}}\int_0^{\pi/2}\int_0^{\pi/2}\frac{\Psi_{4m}}{s_{44m}}\left[\frac{\partial}{\partial\varphi}\left(\frac{p_2\varsigma_{1m}}{\varsigma_m}\right)\frac{\partial}{\partial\varphi}\left(\frac{p_2\varsigma_{2m}}{\varsigma_m}\right)\right.$$

$$\left.+\frac{\partial}{\partial v}\left(\frac{p_2\varsigma_{1m}}{\varsigma_m}\right)\frac{\partial}{\partial v}\left(\frac{p_2\varsigma_{2m}}{\varsigma_m}\right)\right]d\varphi\,dv$$

$$+\frac{4}{s_{44m}}\int_0^{\pi/2}\int_0^{\pi/2}\frac{\Psi_{5m}}{s_{44m}}\left[\frac{\partial}{\partial\varphi}\left(\frac{p_2\varsigma_{1m}}{\varsigma_m}\right)\frac{\partial}{\partial\varphi}\left(\frac{p_2\varsigma_{3m}}{\varsigma_m}\right)\right.$$

$$\left.+\frac{\partial}{\partial v}\left(\frac{p_2\varsigma_{1m}}{\varsigma_m}\right)\frac{\partial}{\partial v}\left(\frac{p_2\varsigma_{3m}}{\varsigma_m}\right)\right]d\varphi\, dv$$

$$+\frac{4}{s_{44m}}\int_0^{\pi/2}\int_0^{\pi/2}\frac{\Psi_{6m}}{s_{44m}}\left[\frac{\partial}{\partial\varphi}\left(\frac{p_2\varsigma_{2m}}{\varsigma_m}\right)\frac{\partial}{\partial\varphi}\left(\frac{p_2\varsigma_{3m}}{\varsigma_m}\right)\right.$$

$$\left.+\frac{\partial}{\partial v}\left(\frac{p_2\varsigma_{2m}}{\varsigma_m}\right)\frac{\partial}{\partial v}\left(\frac{p_2\varsigma_{3m}}{\varsigma_m}\right)\right]d\varphi\, dv\,,$$

$$\sigma_{1B}^{'}=-\sum_{i=1}^{3}\frac{\rho_{iB}p_2\varsigma_{im}}{\varsigma_m}+\frac{1}{s_{44m}}\left[\rho_{1B}^{(\varphi)}\frac{\partial}{\partial\varphi}\left(\frac{p_2\varsigma_{im}}{\varsigma_m}\right)+\rho_{1B}^{(v)}\frac{\partial}{\partial v}\left(\frac{p_2\varsigma_{im}}{\varsigma_m}\right)\right],$$

$$\rho_m=\frac{1}{\varsigma_m}\left[\varsigma_{1m}\left(\frac{4}{3}-\ln R_2\right)+\varsigma_{2m}R_2^{c_{3m}-1}+\frac{\varsigma_{3m}}{R_2}\left(\frac{1}{2}-\ln R_2\right)\right],$$

$$\varsigma_{1m}=r_S^{c_{3m}-1}\left[c_{3m}\left(\frac{1}{2}+\ln r_S\right)-1\right],$$

$$\varsigma_{2m}=\frac{4}{3}-\ln r_S-\left(\frac{1}{2}+\ln r_S\right)\left(\frac{1}{3}-\ln r_S\right),$$

$$\varsigma_{3m}=r_S^{c_{3m}}\left[\frac{1}{3}-\ln r_S-c_{3m}\left(\frac{4}{3}-\ln r_S\right)\right],$$

$$\varsigma_m=r_S^{c_{3m}-1}\left[\frac{c_{1m}-7c_{2m}}{3}-(c_{1m}-c_{2m})\ln R_2\right]$$

$$+\left[(c_{1m}+c_{2m})c_{3m}-2c_{2m}\right]\left(\frac{1}{2}+\ln r_S\right)\left(\frac{1}{3}-\ln r_S\right)R_2^{c_{3m}-1}$$

$$+(c_{1m}-2c_{2m}\ln R_2)\left(\frac{4}{3}-\ln r_S\right)\frac{c_{3m}r_S^{c_{3m}}}{R_2}$$

$$-\left\{c_{3m}\left[\frac{c_{1m}-7c_{2m}}{3}-(c_{1m}-c_{2m})\ln R_2\right]\left(\frac{1}{2}+\ln r_S\right)r_S^{c_{3m}-1}\right.$$

$$+\left[(c_{1m}+c_{2m})c_{3m}-2c_{2m}\right]\left(\frac{4}{3}-\ln r_S\right)R_2^{c_{3m}-1}$$

$$+\frac{(c_{1m}-2c_{2m}\ln R_2)r_S^{c3m}}{R_2}\left(\frac{1}{3}-\ln r_S\right)\Bigg\}. \quad (63)$$

Envelope: The integration constants C_{1e}, C_{2e}, C_{3e} (see Equation (52)) are determined by the mathematical boundary conditions (31), (32), which result in the following combinations: $C_{1e} \neq 0$, $C_{2e} \neq 0$, $C_{3e} = 0$; $C_{1e} \neq 0$, $C_{3e} \neq 0$, $C_{2e} = 0$; $C_{2e} \neq 0$, $C_{3e} \neq 0$, $C_{1e} = 0$. Consequently, such a combination exhibits a minimum value of W_T (see Equation (42)).

Condition $C_{1e} \neq 0$, $C_{2e} \neq 0$, $C_{3e} = 0$. From Equations (18), (31), (31), (32), (52)-(56), we get:

$$\varepsilon'_{11e} = -\left[\varsigma_{1e}\left(\frac{1}{3}-\ln r\right) - c_{3e}\varsigma_{2e}r^{c3e-1}\right],$$

$$\varepsilon'_{22e} = \varepsilon'_{33e} = -\left[\varsigma_{1e}\left(\frac{4}{3}-\ln r\right) + \varsigma_{2e}r^{c3e-1}\right],$$

$$\varepsilon'_{12e} = s_{44e}\,\sigma'_{12e} = -\left[\left(\frac{4}{3}-\ln r\right)\frac{\partial\varsigma_{1e}}{\partial\varphi} + r^{c3e-1}\frac{\partial\varsigma_{2e}}{\partial\varphi}\right],$$

$$\varepsilon'_{13e} = s_{44e}\,\sigma'_{13e} = -\left[\left(\frac{4}{3}-\ln r\right)\frac{\partial\varsigma_{1e}}{\partial v} + r^{c3e-1}\frac{\partial\varsigma_{2e}}{\partial v}\right],$$

$$\sigma'_{11e} = -\left\{\varsigma_{1e}\left[\frac{c_{1e}-7c_{2e}}{3} - (c_{1e}-c_{2e})\ln r\right] + \varsigma_{2e}\left[(c_{1e}+c_{2e})c_{3e} - 2c_{2e}\right]r^{c3e-1}\right\},$$

$$\sigma'_{22e} = \sigma'_{33e} = -\left\{\varsigma_{1e}\left[\frac{4c_{1e}-c_{2e}}{3} - (c_{1e}-c_{2e})\ln r\right] + \varsigma_{2e}(c_{1e}-c_{2e}c_{3e})r^{c3e-1}\right\},$$

$$w_e = \kappa_{1e}\varsigma_{1e}^2 + \kappa_{2e}\varsigma_{2e}^2 + \kappa_{4e}\varsigma_{1e}\varsigma_{2e} + \frac{\chi_{1e}}{s_{44e}}\left[\left(\frac{\partial\varsigma_{1e}}{\partial\varphi}\right)^2 + \left(\frac{\partial\varsigma_{1e}}{\partial v}\right)^2\right]$$

$$+\frac{\chi_{2e}}{s_{44e}}\left[\left(\frac{\partial\varsigma_{2e}}{\partial\varphi}\right)^2 + \left(\frac{\partial\varsigma_{2e}}{\partial v}\right)^2\right] + \frac{\chi_{4e}}{s_{44e}}\left(\frac{\partial\varsigma_{1e}}{\partial\varphi}\frac{\partial\varsigma_{2e}}{\partial\varphi} + \frac{\partial\varsigma_{1e}}{\partial v}\frac{\partial\varsigma_{2e}}{\partial v}\right),$$

$$W_e = 4\int_0^{\pi/2}\int_0^{\pi/2}\left(\Phi_{1e}\varsigma_{1e}^2 + \Phi_{2e}\varsigma_{2e}^2 + \Phi_{4e}\varsigma_{1e}\varsigma_{2e}\right)d\varphi\,dv$$

$$+\frac{4}{s_{44e}}\int_0^{\pi/2}\int_0^{\pi/2}\Psi_{1e}\left[\left(\frac{\partial\varsigma_{1e}}{\partial\varphi}\right)^2+\left(\frac{\partial\varsigma_{1e}}{\partial v}\right)^2\right]d\varphi\, dv$$

$$+\frac{4}{s_{44e}}\int_0^{\pi/2}\int_0^{\pi/2}\Psi_{2e}\left[\left(\frac{\partial\varsigma_{2e}}{\partial\varphi}\right)^2+\left(\frac{\partial\varsigma_{2e}}{\partial v}\right)^2\right]d\varphi\, dv$$

$$+\frac{4}{s_{44e}}\int_0^{\pi/2}\int_0^{\pi/2}\Psi_{4e}\left(\frac{\partial\varsigma_{1e}}{\partial\varphi}\frac{\partial\varsigma_{2e}}{\partial\varphi}+\frac{\partial\varsigma_{1e}}{\partial v}\frac{\partial\varsigma_{2e}}{\partial v}\right)d\varphi\, dv\,, \tag{64}$$

where s_{44e}; c_{ie} (i = 1,2,3), $\kappa_{je},\chi_{je},$, Φ_{je}, Ψ_{je} (j = 1,2,4) are given by Equations (16), (11), (54), (56), respectively. The coefficients ς_{ie}, ς_{ije} (i = 1,2), ς_e are derived as:

$$\varsigma_{ie}=p_1\varsigma_{i1e}+p_2\varsigma_{i2e}\ ,\ i=1,2,$$

$$\varsigma_{11e}=\frac{1}{\varsigma_e}\left[\left(c_{1e}+c_{2e}\right)c_{3e}-2c_{2e}\right]R_2^{c_{3e}-1},$$

$$\varsigma_{12e}=-\frac{1}{\varsigma_e}\left[\left(c_{1e}+c_{2e}\right)c_{3e}-2c_{2e}\right]R_1^{c_{3e}-1}$$

$$\varsigma_{21e}=\frac{1}{\varsigma_e}\left[\frac{c_{1e}-7c_{2e}}{3}-\left(c_{1e}-c_{2e}\right)\ln R_2\right],$$

$$\varsigma_{22e}=-\frac{1}{\varsigma_e}\left[\frac{c_{1e}-7c_{2e}}{3}-\left(c_{1e}-c_{2e}\right)\ln R_1\right],$$

$$\varsigma_e=\left[\frac{c_{1e}-7c_{2e}}{3}\left(R_2^{c_{3e}-1}-R_1^{c_{3e}-1}\right)-\left(c_{1e}-c_{2e}\right)\left(R_2^{c_{3e}-1}\ln R_1-R_1^{c_{3e}-1}\ln R_2\right)\right]$$
$$\times\left[\left(c_{1e}+c_{2e}\right)c_{3e}-2c_{2e}\right]. \tag{65}$$

With regard to $(\varepsilon'_{22e})_{r=R1}$ = - $(p_1\rho_{11e}+p_2\rho_{12e})$, $(\varepsilon'_{22e})_{r=R2}$ = - $(p_1\rho_{21e}+p_2\rho_{22e})$, and Equation (65), the coefficient ρ_{ije} (i, j = 1,2) is derived as:

$$\rho_{ije}=\varsigma_{1je}\left(\frac{4}{3}-\ln R_i\right)+\varsigma_{2je}R_i^{c_{3e}-1},\ i,j=1,2. \tag{66}$$

Condition $C_{1e}\neq 0$, $C_{3e}\neq 0$, $C_{2e}=0$. Similarly, from Equations (18), (31), (31), (32), (52)-(56), we get:

$$\varepsilon'_{11e} = -\left[\varsigma_{1e}\left(\frac{1}{3} - \ln r\right) + \frac{\varsigma_{3e}}{r} \right],$$

$$\varepsilon'_{22e} = \varepsilon'_{33e} = -\left[\varsigma_{1e}\left(\frac{4}{3} - \ln r\right) + \frac{\varsigma_{3e}}{r}\left(\frac{1}{2} + \ln r\right) \right],$$

$$\varepsilon'_{12e} = s_{44e}\,\sigma'_{12e} = -\left[\left(\frac{4}{3} - \ln r\right)\frac{\partial \varsigma_{1e}}{\partial \varphi} + \frac{1}{r}\left(\frac{1}{2} + \ln r\right)\frac{\partial \varsigma_{3e}}{\partial \varphi} \right],$$

$$\varepsilon'_{13e} = s_{44e}\,\sigma'_{13e} = -\left[\left(\frac{4}{3} - \ln r\right)\frac{\partial \varsigma_{1e}}{\partial \nu} + \frac{1}{r}\left(\frac{1}{2} + \ln r\right)\frac{\partial \varsigma_{3e}}{\partial \nu} \right],$$

$$\sigma'_{11e} = -\left\{ \varsigma_{1e}\left[\frac{c_{1e} - 7c_{2e}}{3} - (c_{1e} - c_{2e})\ln r \right] - \frac{\varsigma_{3e}(c_{1e} - 2c_{2e}\ln r)}{r} \right\},$$

$$\sigma'_{22e} = \sigma'_{33e} = -\left\{ \varsigma_{1e}\left[\frac{4c_{1e} - c_{2e}}{3} - (c_{1e} - c_{2e})\ln r \right] + \frac{\varsigma_{3e}}{r}\left(\frac{c_{1e} - 2c_{2e}}{2} + c_{1e}\ln r \right) \right\}$$

$$w_e = \kappa_{1e}\varsigma_{1e}^2 + \kappa_{2e}\varsigma_{2e}^2 + \kappa_{4e}\varsigma_{1e}\varsigma_{2e} + \frac{\chi_{1e}}{s_{44e}}\left[\left(\frac{\partial \varsigma_{1e}}{\partial \varphi}\right)^2 + \left(\frac{\partial \varsigma_{1e}}{\partial \nu}\right)^2 \right]$$

$$+ \frac{\chi_{3e}}{s_{44e}}\left[\left(\frac{\partial \varsigma_{3e}}{\partial \varphi}\right)^2 + \left(\frac{\partial \varsigma_{3e}}{\partial \nu}\right)^2 \right] + \frac{\chi_{5e}}{s_{44e}}\left(\frac{\partial \varsigma_{1e}}{\partial \varphi}\frac{\partial \varsigma_{3e}}{\partial \varphi} + \frac{\partial \varsigma_{1e}}{\partial \nu}\frac{\partial \varsigma_{3e}}{\partial \nu} \right),$$

$$W_e = 4\int_0^{\pi/2}\int_0^{\pi/2}\left(\Phi_{1e}\varsigma_{1e}^2 + \Phi_{3e}\varsigma_{2e}^2 + \Phi_{5e}\varsigma_{1e}\varsigma_{2e} \right) d\varphi\, d\nu$$

$$+ \frac{4}{s_{44e}}\int_0^{\pi/2}\int_0^{\pi/2} \Psi_{1e}\left[\left(\frac{\partial \varsigma_{1e}}{\partial \varphi}\right)^2 + \left(\frac{\partial \varsigma_{1e}}{\partial \nu}\right)^2 \right] d\varphi\, d\nu$$

$$+ \frac{4}{s_{44e}}\int_0^{\pi/2}\int_0^{\pi/2} \Psi_{3e}\left[\left(\frac{\partial \varsigma_{3e}}{\partial \varphi}\right)^2 + \left(\frac{\partial \varsigma_{3e}}{\partial \nu}\right)^2 \right] d\varphi\, d\nu$$

$$+ \frac{4}{s_{44e}}\int_0^{\pi/2}\int_0^{\pi/2} \Psi_{5e}\left(\frac{\partial \varsigma_{1e}}{\partial \varphi}\frac{\partial \varsigma_{3e}}{\partial \varphi} + \frac{\partial \varsigma_{1e}}{\partial \nu}\frac{\partial \varsigma_{3e}}{\partial \nu} \right) d\varphi\, d\nu\,,$$

$$\rho_{ije} = \varsigma_{1je}\left(\frac{4}{3} - \ln R_i\right) + \frac{\varsigma_{3je}}{R_i}\left(\frac{1}{2} + \ln R_i\right),\; i, j = 1,2,$$

$$\varsigma_{ie} = p_1 \varsigma_{i1e} + p_2 \varsigma_{i2e} \ , i = 1,2,3$$

$$\varsigma_{11e} = \frac{1}{\varsigma_e R_2}\left(c_{1e} - 2c_{2e} \ln R_2\right), \quad \varsigma_{12e} = -\frac{1}{\varsigma_e R_1}\left(c_{1e} - 2c_{2e} \ln R_1\right),$$

$$\varsigma_{31e} = -\frac{1}{\varsigma_e}\left[\frac{c_{1e} - 7c_{2e}}{3} - \left(c_{1e} - c_{2e}\right)\ln R_2\right],$$

$$\varsigma_{32e} = \frac{1}{\varsigma_e}\left[\frac{c_{1e} - 7c_{2e}}{3} - \left(c_{1e} - c_{2e}\right)\ln R_1\right],$$

$$\varsigma_e = \frac{c_{1e} - 7c_{2e}}{3}\left(\frac{c_{1e} - 2c_{2e}\ln R_2}{R_2} - \frac{c_{1e} - 2c_{2e}\ln R_1}{R_1}\right)$$

$$-\left(c_{1e} - c_{2e}\right)\left[\frac{\left(c_{1e} - 2c_{2e}\ln R_2\right)\ln R_1}{R_2} - \frac{\left(c_{1e} - 2c_{2e}\ln R_1\right)\ln R_2}{R_1}\right]. \qquad (67)$$

Condition $C_{2e} \neq 0$, $C_{3e} \neq 0$, $C_{1e} = 0$. Similarly, from Equations (18), (31), (31), (32), (52)-(56), we get:

$$\varepsilon'_{11e} = -\left(c_{3e}\varsigma_{2e} r^{c_{3e}-1} + \frac{\varsigma_{3e}}{r}\right),$$

$$\varepsilon'_{22e} = \varepsilon'_{33e} = -\left[\varsigma_{2e} r^{c_{3e}-1} + \frac{\varsigma_{3e}}{r}\left(\frac{1}{2} + \ln r\right)\right],$$

$$\varepsilon'_{12e} = s_{44e}\sigma'_{12e} = -\left[r^{c_{3e}-1}\frac{\partial\varsigma_{2e}}{\partial\varphi} + \frac{1}{r}\left(\frac{1}{2} + \ln r\right)\frac{\partial\varsigma_{3e}}{\partial\varphi}\right],$$

$$\varepsilon'_{13e} = s_{44e}\sigma'_{13e} = -\left[r^{c_{3e}-1}\frac{\partial\varsigma_{2e}}{\partial\nu} + \frac{1}{r}\left(\frac{1}{2} + \ln r\right)\frac{\partial\varsigma_{3e}}{\partial\nu}\right],$$

$$\sigma'_{11e} = -\left\{\varsigma_{2e}\left[\left(c_{1e} + c_{2e}\right)c_{3e} - 2c_{2e}\right]r^{c_{3e}-1} + \frac{\varsigma_{3e}\left(c_{1e} - 2c_{2e}\ln r\right)}{r}\right\},$$

$$\sigma'_{22e} = \sigma'_{33e} = -\left[\varsigma_{2e}\left(c_{1e} - c_{2e}c_{3e}\right)r^{c_{3e}-1} + \frac{\varsigma_{3e}}{r}\left(\frac{c_{1e} - 2c_{2e}}{2} + c_{1e}\ln r\right)\right],$$

$$w_e = \kappa_{2e}\varsigma_{2e}^2 + \kappa_{3e}\varsigma_{3e}^2 + \kappa_{6e}\varsigma_{2e}\varsigma_{3e} + \frac{\chi_{2e}}{s_{44e}}\left[\left(\frac{\partial\varsigma_{2e}}{\partial\varphi}\right)^2 + \left(\frac{\partial\varsigma_{2e}}{\partial\nu}\right)^2\right]$$

$$+\frac{\chi_{3e}}{s_{44e}}\left[\left(\frac{\partial\varsigma_{3e}}{\partial\varphi}\right)^2+\left(\frac{\partial\varsigma_{3e}}{\partial\nu}\right)^2\right]+\frac{\chi_{6e}}{s_{44e}}\left(\frac{\partial\varsigma_{2e}}{\partial\varphi}\frac{\partial\varsigma_{3e}}{\partial\varphi}+\frac{\partial\varsigma_{2e}}{\partial\nu}\frac{\partial\varsigma_{3e}}{\partial\nu}\right),$$

$$W_e=4\int_0^{\pi/2}\int_0^{\pi/2}\left(\Phi_{2e}\varsigma_{2e}^2+\Phi_{3e}\varsigma_{3e}^2+\Phi_{6e}\varsigma_{2e}\varsigma_{3e}\right)d\varphi\,d\nu$$

$$+\frac{4}{s_{44e}}\int_0^{\pi/2}\int_0^{\pi/2}\Psi_{2e}\left[\left(\frac{\partial\varsigma_{2e}}{\partial\varphi}\right)^2+\left(\frac{\partial\varsigma_{2e}}{\partial\nu}\right)^2\right]d\varphi\,d\nu$$

$$+\frac{4}{s_{44e}}\int_0^{\pi/2}\int_0^{\pi/2}\Psi_{3e}\left[\left(\frac{\partial\varsigma_{3e}}{\partial\varphi}\right)^2+\left(\frac{\partial\varsigma_{3e}}{\partial\nu}\right)^2\right]d\varphi\,d\nu$$

$$+\frac{4}{s_{44e}}\int_0^{\pi/2}\int_0^{\pi/2}\Psi_{6e}\left(\frac{\partial\varsigma_{2e}}{\partial\varphi}\frac{\partial\varsigma_{3e}}{\partial\varphi}+\frac{\partial\varsigma_{2e}}{\partial\nu}\frac{\partial\varsigma_{3e}}{\partial\nu}\right)d\varphi\,d\nu,$$

$$\rho_{ije}=\varsigma_{2je}R_i^{c_{3e}-1}+\frac{\varsigma_{3je}}{R_i}\left(\frac{1}{2}+\ln R_i\right),\ i,j=1,2,$$

$$\varsigma_{ie}=p_1\varsigma_{i1e}+p_2\varsigma_{i2e},\ i=2,3,$$

$$\varsigma_{21e}=\frac{1}{\varsigma_e R_2}\left(c_{1e}-2c_{2e}\ln R_2\right),\quad \varsigma_{22e}=-\frac{1}{\varsigma_e R_1}\left(c_{1e}-2c_{2e}\ln R_1\right),$$

$$\varsigma_{31e}=-\frac{1}{\varsigma_e}\left[\left(c_{1e}+c_{2e}\right)c_{3e}-2c_{2e}\right]R_2^{c_{3e}-1},$$

$$\varsigma_{32e}=\frac{1}{\varsigma_e}\left[\left(c_{1e}+c_{2e}\right)c_{3e}-2c_{2e}\right]R_1^{c_{3e}-1},$$

$$\varsigma_e=\left[\left(c_{1e}+c_{2e}\right)c_{3e}-2c_{2e}\right]$$
$$\times\left[\frac{\left(c_{1e}-2c_{2e}\ln R_2\right)R_1^{c_{3e}-1}}{R_2}-\frac{\left(c_{1e}-2c_{2e}\ln R_1\right)R_2^{c_{3e}-1}}{R_1}\right] \quad (68)$$

Let the elastic energy $W_{eB}=W_{ieB}$ be related to the condition $C_{ieB}\neq C_{jeB}=C_{keB}$, ($i$, j, k = 1,2,3; $i\neq j\neq k$). If $W_{ieB}<W_{jeB}$ and $W_{ieB}<W_{keB}$, then the mathematical solution for $C_{ieB}\neq C_{jeB}=C_{keB}$ is considered to result in a minimum value of W_T (see Equation (42)).

Condition $C_{1eB} \neq C_{2eB} = C_{3eB} = 0$. With regard to Equations (18), (31), (45), (52)-(56), we get:

$$\varepsilon'_{11eB} = -\frac{\sigma'_{1B}\rho_{mB}}{\varsigma_{eB}}\left(\frac{1}{3} - \ln r\right),$$

$$\varepsilon'_{22eB} = \varepsilon'_{33eB} = -\frac{\sigma'_{1B}\rho_{mB}}{\varsigma_{eB}}\left(\frac{4}{3} - \ln r\right),$$

$$\varepsilon'_{12eB} = s_{44e}\,\sigma'_{12eB} = -\left(\frac{4}{3} - \ln r\right)\frac{\partial}{\partial\varphi}\left(\frac{\sigma'_{1B}\rho_{mB}}{\varsigma_{eB}}\right),$$

$$\varepsilon'_{13eB} = s_{44e}\,\sigma'_{13eB} = -\left(\frac{4}{3} - \ln r\right)\frac{\partial}{\partial v}\left(\frac{\sigma'_{1B}\rho_{mB}}{\varsigma_{eB}}\right),$$

$$\sigma'_{11eB} = -\frac{\sigma'_{1B}\rho_{mB}}{\varsigma_{eB}}\left[\frac{c_{1e} - 7c_{2e}}{3} - (c_{1e} - c_{2e})\ln r\right],$$

$$\sigma'_{22eB} = \sigma'_{33eB} = -\frac{\sigma'_{1B}\rho_{mB}}{\varsigma_{eB}}\left[\frac{4c_{1e} - c_{2e}}{3} - (c_{1e} - c_{2e})\ln r\right],$$

$$w_{eB} = \kappa_{1e}\left(\frac{\sigma'_{1B}\rho_{mB}}{\varsigma_{eB}}\right)^2 + \frac{\chi_{1e}}{s_{44e}}\left\{\left[\frac{\partial}{\partial\varphi}\left(\frac{\sigma'_{1B}\rho_{mB}}{\varsigma_{eB}}\right)\right]^2 + \left[\frac{\partial}{\partial v}\left(\frac{\sigma'_{1B}\rho_{mB}}{\varsigma_{eB}}\right)\right]^2\right\}$$

$$W_{eB} = 4\int_0^{\pi/2}\int_0^{\pi/2}\Phi_{1e}\left(\frac{\sigma'_{1B}\rho_{mB}}{\varsigma_{eB}}\right)^2 d\varphi\, dv$$

$$+\frac{4}{s_{44e}}\int_0^{\pi/2}\int_0^{\pi/2}\Psi_{1e}\left\{\left[\frac{\partial}{\partial\varphi}\left(\frac{\sigma'_{1B}\rho_{mB}}{\varsigma_{eB}}\right)\right]^2 + \left[\frac{\partial}{\partial v}\left(\frac{\sigma'_{1B}\rho_{mB}}{\varsigma_{eB}}\right)\right]^2\right\} d\varphi\, dv, \quad (69)$$

where ρ_{mB} is given by Equations (60)-(62), and the coefficient ς_{eB} is derived as:

$$\varsigma_{eB} = \frac{4}{3} - \ln R_2. \quad (70)$$

From $(\varepsilon'_{22eB})_{r=R1} = - \sigma'_{1B}\, \rho_{eB}$ and Equation (88), the coefficient ρ_{eB} is derived as:

$$\rho_{eB} = \frac{\rho_{mB}}{\varsigma_{eB}}\left(\frac{4}{3} - \ln R_1\right). \tag{71}$$

Condition $C_{2eB} \neq C_{1eB} = C_{3eB} = 0$. Similarly, with regard to Equations (18), (31), (45), (52)-(56), we get:

$$\varepsilon'_{11eB} = -\sigma'_{1B}\rho_{mB}\, c_{3e}\left(\frac{r}{R_2}\right)^{c3e-1},$$

$$\varepsilon'_{22eB} = \varepsilon'_{33eB} = -\sigma'_{1B}\rho_{mB}\left(\frac{r}{R_2}\right)^{c3e-1},$$

$$\varepsilon'_{12eB} = s_{44e}\,\sigma'_{12eB} = -r^{c3e-1}\frac{\partial}{\partial\varphi}\left(\frac{\sigma'_{1B}\rho_{mB}}{R_2^{c3e-1}}\right),$$

$$\varepsilon'_{13eB} = s_{44e}\,\sigma'_{13eB} = -r^{c3e-1}\frac{\partial}{\partial\nu}\left(\frac{\sigma'_{1B}\rho_{mB}}{R_2^{c3e-1}}\right),$$

$$\sigma'_{11eB} = -\sigma'_{1B}\rho_{mB}\left[\left(c_{1e} + c_{2e}\right)c_{3e} - 2c_{2e}\right]\left(\frac{r}{R_2}\right)^{c3e-1},$$

$$\sigma'_{22eB} = \sigma'_{33eB} = -\sigma'_{1B}\rho_{mB}\left(c_{1e} - c_{2e}\,c_{3e}\right)\left(\frac{r}{R_2}\right)^{c3e-1},$$

$$w_{eB} = \kappa_{2e}\left(\frac{\sigma'_{1B}\rho_{mB}}{R_2^{c3e-1}}\right)^2 + \frac{\chi_{2e}}{s_{44e}}\left\{\left[\frac{\partial}{\partial\varphi}\left(\frac{\sigma'_{1B}\rho_{mB}}{R_2^{c3e-1}}\right)\right]^2 + \left[\frac{\partial}{\partial\nu}\left(\frac{\sigma'_{1B}\rho_{mB}}{R_2^{c3e-1}}\right)\right]^2\right\},$$

$$W_{eB} = 4\int_0^{\pi/2}\int_0^{\pi/2}\Phi_{2e}\left(\frac{\sigma'_{1B}\rho_{mB}}{R_2^{c3e-1}}\right)^2 d\varphi\, d\nu$$

$$+\frac{4}{s_{44e}}\int_0^{\pi/2}\int_0^{\pi/2}\Psi_{2e}\left\{\left[\frac{\partial}{\partial\varphi}\left(\frac{\sigma'_{1B}\rho_{mB}}{R_2^{c3e-1}}\right)\right]^2+\left[\frac{\partial}{\partial v}\left(\frac{\sigma'_{1B}\rho_{mB}}{R_2^{c3e-1}}\right)\right]^2\right\}d\varphi\, dv,$$

$$\rho_{eB}=\rho_{mB}\left(\frac{R_1}{R_2}\right)^{c3e-1}. \tag{72}$$

Condition $C_{3eB}\neq C_{1eB}=C_{2eB}=0$. Finally, from Equations (18), (31), (45), (52)-(56), we get:

$$\varepsilon'_{11eB}=-\frac{\sigma'_{1B}\rho_{mB}}{\varsigma_{eB}r},\ \varepsilon'_{22eB}=\varepsilon'_{33eB}=-\frac{\sigma'_{1B}\rho_{mB}}{\varsigma_{eB}r}\left(\frac{1}{2}+\ln r\right),$$

$$\varepsilon'_{12eB}=s_{44e}\sigma'_{12eB}=-\frac{1}{r}\left(\frac{1}{2}+\ln r\right)\frac{\partial}{\partial\varphi}\left(\frac{\sigma'_{1B}\rho_{mB}}{\varsigma_{eB}}\right),$$

$$\varepsilon'_{13eB}=s_{44e}\sigma'_{13eB}=-\frac{1}{r}\left(\frac{1}{2}+\ln r\right)\frac{\partial}{\partial v}\left(\frac{\sigma'_{1B}\rho_{mB}}{\varsigma_{eB}}\right),$$

$$\sigma'_{11eB}=-\frac{\sigma'_{1B}\rho_{mB}}{\varsigma_{eB}r}\left(c_{1e}-2c_{2e}\ln r\right),$$

$$\sigma'_{22eB}=\sigma'_{33eB}=-\frac{\sigma'_{1B}\rho_{mB}}{\varsigma_{eB}r}\left(\frac{c_{1e}-2c_{2e}}{2}+c_{1e}\ln r\right),$$

$$w_{eB}=\kappa_{3e}\left(\frac{\sigma'_{1B}\rho_{mB}}{\varsigma_{eB}}\right)^2+\frac{\chi_{3e}}{s_{44e}}\left\{\left[\frac{\partial}{\partial\varphi}\left(\frac{\sigma'_{1B}\rho_{mB}}{\varsigma_{eB}}\right)\right]^2+\left[\frac{\partial}{\partial v}\left(\frac{\sigma'_{1B}\rho_{mB}}{\varsigma_{eB}}\right)\right]^2\right\},$$

$$W_{eB}=4\int_0^{\pi/2}\int_0^{\pi/2}\Phi_{3e}\left(\frac{\sigma'_{1B}\rho_{mB}}{\varsigma_{eB}}\right)^2 d\varphi\, dv$$

$$+\frac{4}{s_{44e}}\int_0^{\pi/2}\int_0^{\pi/2}\Psi_{3e}\left\{\left[\frac{\partial}{\partial\varphi}\left(\frac{\sigma'_{1B}\rho_{mB}}{\varsigma_{eB}}\right)\right]^2+\left[\frac{\partial}{\partial v}\left(\frac{\sigma'_{1B}\rho_{mB}}{\varsigma_{eB}}\right)\right]^2\right\}d\varphi\, dv,$$

$$\varsigma_{eB}=\frac{4}{3}-\ln R_2,\ \rho_{eB}=\frac{\rho_{mB}}{\varsigma_{eB}R_1}\left(\frac{1}{2}+\ln R_1\right). \tag{73}$$

Particle: The mathematical solution for the radial displacement u'_{1p} in the spherical particle has the form:

$$u'_{1p} = C_{1p} r^{\lambda p}, \lambda_p = \frac{1}{2}\left[1 + \sqrt{1 + 16(1 - \mu_p)[1 + 4(1 - \mu_p)]}\right] > 3, \quad (74)$$

where $\mu_p < 0.5$ for an accurate isotropic material. The subscript q = p, pB and the stress σ_q in this section are derived as:

$$q = p \Rightarrow \sigma_q = p_1, \quad \rho_q = \rho_p = \frac{1}{\xi_{1p}} = \frac{(1 + \mu_p)(1 - 2\mu_p)}{E_p[\lambda_{1p}(1 + \mu_p) + 2\mu_p]},$$

$$q = mB \Rightarrow \sigma_q = \rho_{eB}, \quad (75)$$

where p_1, r_{eB}, and ξ_{1p} are given by Equations (24), (71)-(73), and (75), respectively. From Equations (3)-(6), (7)-(10), (18), (19), (28), (46), (74), we get:

$$\varepsilon'_{11q} = -\sigma_q \rho_q \lambda_{1p} \left(\frac{r}{R_1}\right)^{\lambda 1p-1},$$

$$\varepsilon'_{22q} = \varepsilon'_{33q} = -\sigma_q \rho_q \left(\frac{r}{R_1}\right)^{\lambda 1p-1},$$

$$\varepsilon'_{12q} = s_{44p}\sigma'_{12q} = -\rho_q \left(\frac{r}{R_1}\right)^{\lambda 1p-1} \frac{\partial \sigma_q}{\partial \varphi},$$

$$\varepsilon'_{13q} = s_{44p}\sigma'_{13q} = -\rho_q \left(\frac{r}{R_1}\right)^{\lambda 1p-1} \frac{\partial \sigma_q}{\partial \nu},$$

$$\sigma'_{11q} = -\sigma_q \rho_q \xi_{1p} \left(\frac{r}{R_1}\right)^{\lambda 1p-1},$$

$$\sigma'_{22q} = \sigma'_{33q} = -\sigma_q \rho_q \xi_{2p} \left(\frac{r}{R_1}\right)^{\lambda 1p-1},$$

$$w_q = \kappa_q \left(\frac{r}{R_1}\right)^{2(\lambda 1p-1)}, W_q = \frac{4R_1^3}{2\lambda_{1p}+1} \int_0^{\pi/2} \int_0^{\pi/2} \kappa_q \, d\varphi \, d\nu, \quad (76)$$

where ξ_i ($i = 1,2,3$), κ_q are derived as:

$$\xi_{1p} = \frac{E_p\left[\lambda_{1p}\left(1+\mu_p\right)+2\mu_p\right]}{\left(1+\mu_p\right)\left(1-2\mu_p\right)},\ \xi_{2p} = \frac{E_p\left[\lambda_{1p}\left(1+\mu_p\right)+2\mu_p\right]}{\left(1+\mu_p\right)\left(1-2\mu_p\right)},$$

$$\xi_{3p} = \frac{E_p\left\{\lambda_{1p}\left[\lambda_{1p}\left(1-\mu_p\right)+4\mu_p\right]+2\right\}}{\left(1+\mu_p\right)\left(1-2\mu_p\right)},$$

$$\kappa_p = \xi_{3p}\left(\sigma_q\,\rho_q\right)^2 + \frac{1}{s_{44p}}\left\{\left[\frac{\partial\left(\sigma_q\,\rho_q\right)}{\partial\varphi}\right]^2 + \left[\frac{\partial\left(\sigma_q\,\rho_q\right)}{\partial\nu}\right]^2\right\}. \tag{77}$$

Condition $\beta_p \neq \beta_c = \beta_m$

Matrix: The thermal stresses in the cell matrix for the subscript mB are given by Equations (58), (61), and (62). The integration constants C_{1m}, C_{2m}, C_{3m} (see Equation (52)) are determined by the mathematical boundary conditions (34), (35) or (34)-(36), which result in the following combinations: $C_{1m} \neq 0$, $C_{2m} \neq 0$, $C_{3m} = 0$; $C_{1m} \neq 0$, $C_{3m} \neq 0$, $C_{2m} = 0$; $C_{2m} \neq 0$, $C_{3m} \neq 0$, $C_{1m} = 0$; or $C_{1m} \neq 0$, $C_{2m} \neq 0$, $C_{3m} \neq 0$, respectively. Consequently, such a combination exhibits a minimum value of W_T (see Equation (42)).

Condition $C_{1m} \neq 0$, $C_{2m} \neq 0$, $C_{3m} = 0$. From Equations (18), (32), (34), (35), (52)-(56), we get:

$$\varepsilon'_{11m} = \varsigma_{1m}\left(\frac{1}{3} - \ln r\right) + \varsigma_{2m} c_{3m} r^{c_{3m}-1},$$

$$\varepsilon'_{22m} = \varepsilon'_{33m} = \varsigma_{1m}\left(\frac{4}{3} - \ln r\right) + \varsigma_{2m} r^{c_{3m}-1},$$

$$\varepsilon'_{12m} = s_{44m}\,\sigma'_{12m} = \left(\frac{4}{3} - \ln r\right)\frac{\partial\varsigma_{1m}}{\partial\varphi} + r^{c_{3m}-1}\frac{\partial\varsigma_{2m}}{\partial\varphi},$$

$$\varepsilon'_{13m} = s_{44m}\,\sigma'_{13m} = \left(\frac{4}{3} - \ln r\right)\frac{\partial\varsigma_{1m}}{\partial\nu} + r^{c_{3m}-1}\frac{\partial\varsigma_{2m}}{\partial\nu},$$

$$\sigma'_{11m} = -\,\varsigma_{1m}\left[\frac{c_{1m} - 7c_{2m}}{3} - \left(c_{1m} - c_{2m}\right)\ln r\right]$$

$$+\,\varsigma_{2m}\left[\left(c_{1m} + c_{2m}\right)c_{3m} - 2c_{2m}\right] r^{c_{3m}-1},$$

$$\sigma'_{22m} = \sigma'_{33m} = \varsigma_{1m}\left[\frac{4c_{1m} - c_{2m}}{3} - (c_{1m} - c_{2m})\ln r\right] + \varsigma_{2m}(c_{1m} - c_{2m}c_{3m})r^{c_{3m}-1}$$

$$w_m = \kappa_{1m}\varsigma_{1m}^2 + \kappa_{2m}\varsigma_{2m}^2 + \kappa_{4m}\varsigma_{1m}\varsigma_{2m} + \frac{\chi_{1m}}{s_{44m}}\left[\left(\frac{\partial\varsigma_{1m}}{\partial\varphi}\right)^2 + \left(\frac{\partial\varsigma_{1m}}{\partial v}\right)^2\right]$$

$$+ \frac{\chi_{2m}}{s_{44m}}\left[\left(\frac{\partial\varsigma_{2m}}{\partial\varphi}\right)^2 + \left(\frac{\partial\varsigma_{2m}}{\partial v}\right)^2\right] + + \frac{\chi_{4m}}{s_{44m}}\left(\frac{\partial\varsigma_{1m}}{\partial\varphi}\frac{\partial\varsigma_{2m}}{\partial\varphi} + \frac{\partial\varsigma_{1m}}{\partial v}\frac{\partial\varsigma_{2m}}{\partial v}\right),$$

$$W_m = 4\int_0^{\pi/2}\int_0^{\pi/2}\left(\Phi_{1m}\varsigma_{1m}^2 + \Phi_{2m}\varsigma_{2m}^2 + \Phi_{4m}\varsigma_{1m}\varsigma_{2m}\right) d\varphi\, dv$$

$$+ \frac{4}{s_{44m}}\int_0^{\pi/2}\int_0^{\pi/2}\frac{\Psi_{1m}}{s_{44m}}\left[\left(\frac{\partial\varsigma_{1m}}{\partial\varphi}\right)^2 + \left(\frac{\partial\varsigma_{1m}}{\partial v}\right)^2\right] d\varphi\, dv$$

$$+ \frac{4}{s_{44m}}\int_0^{\pi/2}\int_0^{\pi/2}\frac{\Psi_{2m}}{s_{44m}}\left[\left(\frac{\partial\varsigma_{2m}}{\partial\varphi}\right)^2 + \left(\frac{\partial\varsigma_{2m}}{\partial v}\right)^2\right] d\varphi\, dv$$

$$+ \frac{4}{s_{44m}}\int_0^{\pi/2}\int_0^{\pi/2}\frac{\Psi_{4m}}{s_{44m}}\left(\frac{\partial\varsigma_{1m}}{\partial\varphi}\frac{\partial\varsigma_{2m}}{\partial\varphi} + \frac{\partial\varsigma_{1m}}{\partial v}\frac{\partial\varsigma_{2m}}{\partial v}\right) d\varphi\, dv, \quad (78)$$

where r_S, s_{44m}, c_{im} ($i = 1,2,3$), $\kappa_{jm}, \chi_{jm}, \Phi_{jm}, \Psi_{jm}$ ($j = 1,2,4$) are given by Equations (2), (16), (11), (54), (56), respectively. The coefficients ς_m, ς_{im} ($i = 1,2$) are derived as:

$$\varsigma_{1m} = -\frac{p_1\,\rho_{2me}\,R_2\,r_S^{c_{3m}}}{\varsigma_m},$$

$$\varsigma_{2m} = \frac{p_1\,\rho_{2me}\,R_2\,r_S^{c_{3m}}}{\varsigma_m}\left(\frac{4}{3} - \ln r_S\right),$$

$$\varsigma_m = r_S^{c_{3m}} R_2\left(\frac{4}{3} - \ln R_2\right) - R_2^{c_{3m}} r_S\left(\frac{4}{3} - \ln r_S\right), \quad (79)$$

where p_2 and ρ_{2me} are given by Equations (26) and (75), respectively.

Condition $C_{1m} \neq 0$, $C_{3m} \neq 0$, $C_{2m} = 0$. From Equations (18), (32), (34), (35), (52)-(56), we get:

$$\varepsilon'_{11m} = \varsigma_{1m}\left(\frac{1}{3} - \ln r\right) + \frac{\varsigma_{3m}}{r},$$

$$\varepsilon'_{22m} = \varepsilon'_{33m} = \varsigma_{1m}\left(\frac{4}{3} - \ln r\right) + \frac{\varsigma_{3m}}{r}\left(\frac{1}{2} + \ln r\right),$$

$$\varepsilon'_{12m} = s_{44m}\,\sigma'_{12m} = \left(\frac{4}{3} - \ln r\right)\frac{\partial \varsigma_{1m}}{\partial \varphi} + \frac{1}{r}\left(\frac{1}{2} + \ln r\right)\frac{\partial \varsigma_{3m}}{\partial \varphi},$$

$$\varepsilon'_{13m} = s_{44m}\,\sigma'_{13m} = \left(\frac{4}{3} - \ln r\right)\frac{\partial \varsigma_{1m}}{\partial \nu} + \frac{1}{r}\left(\frac{1}{2} + \ln r\right)\frac{\partial \varsigma_{3m}}{\partial \nu},$$

$$\sigma'_{11m} = \varsigma_{1m}\left[\frac{c_{1m} - 7c_{2m}}{3} - (c_{1m} - c_{2m})\ln r\right] + \frac{\varsigma_{3m}(c_{1m} - 2c_{2m}\ln r)}{r},$$

$$\sigma'_{22m} = \sigma'_{33m} = \varsigma_{1m}\left[\frac{4c_{1m} - c_{2m}}{3} - (c_{1m} - c_{2m})\ln r\right]$$

$$+ \frac{\varsigma_{3m}}{r}\left(\frac{c_{1m} - 2c_{2m}}{3} + c_{1m}\ln r\right),$$

$$w_m = \kappa_{1m}\varsigma_{1m}^2 + \kappa_{2m}\varsigma_{2m}^2 + \kappa_{5m}\varsigma_{1m}\varsigma_{3m} + \frac{\chi_{1m}}{s_{44m}}\left[\left(\frac{\partial \varsigma_{1m}}{\partial \varphi}\right)^2 + \left(\frac{\partial \varsigma_{1m}}{\partial \nu}\right)^2\right]$$

$$+ \frac{\chi_{3m}}{s_{44m}}\left[\left(\frac{\partial \varsigma_{3m}}{\partial \varphi}\right)^2 + \left(\frac{\partial \varsigma_{3m}}{\partial \nu}\right)^2\right] + \frac{\chi_{5m}}{s_{44m}}\left(\frac{\partial \varsigma_{1m}}{\partial \varphi}\frac{\partial \varsigma_{3m}}{\partial \varphi} + \frac{\partial \varsigma_{1m}}{\partial \nu}\frac{\partial \varsigma_{3m}}{\partial \nu}\right),$$

$$W_m = 4\int_0^{\pi/2}\int_0^{\pi/2}\left(\Phi_{1m}\varsigma_{1m}^2 + \Phi_{3m}\varsigma_{2m}^2 + \Phi_{5m}\varsigma_{1m}\varsigma_{3m}\right) d\varphi\, d\nu$$

$$+ \frac{4}{s_{44m}}\int_0^{\pi/2}\int_0^{\pi/2}\frac{\Psi_{1m}}{s_{44m}}\left[\left(\frac{\partial \varsigma_{1m}}{\partial \varphi}\right)^2 + \left(\frac{\partial \varsigma_{1m}}{\partial \nu}\right)^2\right] d\varphi\, d\nu$$

$$+ \frac{4}{s_{44m}}\int_0^{\pi/2}\int_0^{\pi/2}\frac{\Psi_{3m}}{s_{44m}}\left[\left(\frac{\partial \varsigma_{3m}}{\partial \varphi}\right)^2 + \left(\frac{\partial \varsigma_{3m}}{\partial \nu}\right)^2\right] d\varphi\, d\nu$$

$$+\frac{4}{s_{44m}}\int_0^{\pi/2}\int_0^{\pi/2}\frac{\Psi_{5m}}{s_{44m}}\left(\frac{\partial\varsigma_{1m}}{\partial\varphi}\frac{\partial\varsigma_{3m}}{\partial\varphi}+\frac{\partial\varsigma_{1m}}{\partial\nu}\frac{\partial\varsigma_{3m}}{\partial\nu}\right)d\varphi\, d\nu\,,$$

$$\varsigma_{1m}=-\frac{p_1\,\rho_{2me}\,R_2}{\varsigma_m}\left(\frac{1}{2}+\ln r_S\right),$$

$$\varsigma_{3m}=\frac{p_1\,\rho_{2me}\,R_2\,r_S}{\varsigma_m}\left(\frac{4}{3}-\ln r_S\right),$$

$$\varsigma_m=R_2\left(\frac{4}{3}-\ln R_2\right)\left(\frac{1}{2}+\ln r_S\right)-r_S\left(\frac{4}{3}-\ln r_S\right)\left(\frac{1}{2}+\ln R_2\right). \tag{80}$$

Condition $C_{2m}\neq 0$, $C_{3m}\neq 0$, $C_{1m}=0$. From Equations (18), (32), (34), (35), (52)-(56), we get:

$$\varepsilon'_{11m}=\varsigma_{2m}\,c_{3m}\,r^{c_{3m}-1}+\frac{\varsigma_{3m}}{r}\,,$$

$$\varepsilon'_{22m}=\varepsilon'_{33m}=\varsigma_{2m}\;r^{c_{3m}-1}+\frac{\varsigma_{3m}}{r}\left(\frac{1}{2}+\ln r\right),$$

$$\varepsilon'_{12m}=s_{44m}\,\sigma'_{12m}=r^{c_{3m}-1}\frac{\partial\varsigma_{2m}}{\partial\varphi}+\frac{1}{r}\left(\frac{1}{2}+\ln r\right)\frac{\partial\varsigma_{3m}}{\partial\varphi}\,,$$

$$\varepsilon'_{13m}=s_{44m}\,\sigma'_{13m}=r^{c_{3m}-1}\frac{\partial\varsigma_{2m}}{\partial\nu}+\frac{1}{r}\left(\frac{1}{2}+\ln r\right)\frac{\partial\varsigma_{3m}}{\partial\nu}\,,$$

$$\sigma'_{11m}=\varsigma_{2m}\left[\left(c_{1m}+c_{2m}\right)c_{3m}-2c_{2m}\right]r^{c_{3m}-1}+\frac{\varsigma_{3m}\left(c_{1m}-2c_{2m}\ln r\right)}{r}\,,$$

$$\sigma'_{22m}=\sigma'_{33m}=\varsigma_{2m}\left(c_{1m}-c_{2m}c_{3m}\right)r^{c_{3m}-1}+\frac{\varsigma_{3m}}{r}\left(\frac{c_{1m}-2c_{2m}}{3}+c_{1m}\ln r\right),$$

$$w_m=\kappa_{2m}\varsigma_{2m}^2+\kappa_{3m}\varsigma_{3m}^2+\kappa_{6m}\varsigma_{2m}\varsigma_{3m}+\frac{\chi_{2m}}{s_{44m}}\left[\left(\frac{\partial\varsigma_{2m}}{\partial\varphi}\right)^2+\left(\frac{\partial\varsigma_{2m}}{\partial\nu}\right)^2\right]$$

$$+\frac{\chi_{3m}}{s_{44m}}\left[\left(\frac{\partial\varsigma_{3m}}{\partial\varphi}\right)^2+\left(\frac{\partial\varsigma_{3m}}{\partial\nu}\right)^2\right]+\frac{\chi_{6m}}{s_{44m}}\left(\frac{\partial\varsigma_{2m}}{\partial\varphi}\frac{\partial\varsigma_{3m}}{\partial\varphi}+\frac{\partial\varsigma_{2m}}{\partial\nu}\frac{\partial\varsigma_{3m}}{\partial\nu}\right),$$

$$W_m=4\int_0^{\pi/2}\int_0^{\pi/2}\left(\Phi_{2m}\varsigma_{2m}^2+\Phi_{3m}\varsigma_{3m}^2+\Phi_{6m}\varsigma_{2m}\varsigma_{3m}\right)d\varphi\, d\nu$$

$$+\frac{4}{s_{44m}}\int_0^{\pi/2}\int_0^{\pi/2}\frac{\Psi_{2m}}{s_{44m}}\left[\left(\frac{\partial\varsigma_{2m}}{\partial\varphi}\right)^2+\left(\frac{\partial\varsigma_{2m}}{\partial v}\right)^2\right]d\varphi\,dv$$

$$+\frac{4}{s_{44m}}\int_0^{\pi/2}\int_0^{\pi/2}\frac{\Psi_{3m}}{s_{44m}}\left[\left(\frac{\partial\varsigma_{3m}}{\partial\varphi}\right)^2+\left(\frac{\partial\varsigma_{3m}}{\partial v}\right)^2\right]d\varphi\,dv$$

$$+\frac{4}{s_{44m}}\int_0^{\pi/2}\int_0^{\pi/2}\frac{\Psi_{6m}}{s_{44m}}\left(\frac{\partial\varsigma_{2m}}{\partial\varphi}\frac{\partial\varsigma_{3m}}{\partial\varphi}+\frac{\partial\varsigma_{2m}}{\partial v}\frac{\partial\varsigma_{3m}}{\partial v}\right)d\varphi\,dv\,,$$

$$\varsigma_{1m}=-\frac{p_1\,\rho_{2me}\,R_2}{\varsigma_m}\left(\frac{1}{2}+\ln r_S\right),$$

$$\varsigma_{2m}=\frac{p_1\,\rho_{2me}\,R_2\,r_S^{c_{3m}}}{\varsigma_m},$$

$$\varsigma_m=R_2^{c_{3m}}\left(\frac{1}{2}+\ln r_S\right)-r_S^{c_{3m}}\left(\frac{1}{2}+\ln R_2\right). \tag{81}$$

Condition $C_{1m}\neq 0$, $C_{2m}\neq 0$, $C_{3m}\neq 0$. From Equation (18), (32), (34), (35), (52)-(56), we get:

$$\varepsilon'_{11m}=\varsigma_{1m}\left(\frac{1}{3}-\ln r\right)+\varsigma_{2m}\,c_{3m}\,r^{c_{3m}-1}+\frac{\varsigma_{3m}}{r},$$

$$\varepsilon'_{22m}=\varepsilon'_{33m}=\varsigma_{1m}\left(\frac{4}{3}-\ln r\right)+\varsigma_{2m}\;r^{c_{3m}-1}+\frac{\varsigma_{3m}}{r}\left(\frac{1}{2}+\ln r\right),$$

$$\varepsilon'_{12m}=s_{44m}\,\sigma'_{12m}=\left(\frac{4}{3}-\ln r\right)\frac{\partial\varsigma_{1m}}{\partial\varphi}+r^{c_{3m}-1}\frac{\partial\varsigma_{2m}}{\partial\varphi}+\frac{1}{r}\left(\frac{1}{2}+\ln r\right)\frac{\partial\varsigma_{3m}}{\partial\varphi},$$

$$\varepsilon'_{13m}=s_{44m}\,\sigma'_{13m}=\left(\frac{4}{3}-\ln r\right)\frac{\partial\varsigma_{1m}}{\partial v}+r^{c_{3m}-1}\frac{\partial\varsigma_{2m}}{\partial v}+\frac{1}{r}\left(\frac{1}{2}+\ln r\right)\frac{\partial\varsigma_{3m}}{\partial v},$$

$$\sigma'_{11m}=\varsigma_{1m}\left[\frac{c_{1m}-7c_{2m}}{3}-\left(c_{1m}-c_{2m}\right)\ln r\right]$$

$$+\varsigma_{2m}\left[\left(c_{1m}+c_{2m}\right)c_{3m}-2c_{2m}\right]r^{c_{3m}-1}+\frac{\varsigma_{3m}\left(c_{1m}-2c_{2m}\ln r\right)}{r},$$

$$\sigma'_{22m}=\sigma'_{33m}=\varsigma_{1m}\left[\frac{4c_{1m}-c_{2m}}{3}-\left(c_{1m}-c_{2m}\right)\ln r\right]$$

$$+\varsigma_{2m}\left(c_{1m}-c_{2m}c_{3m}\right)r^{c_{3m}-1}+\frac{\varsigma_{3m}}{r}\left(\frac{c_{1m}-2c_{2m}}{3}+c_{1m}\ln r\right),$$

$$w_m=\kappa_{1m}\varsigma_{1m}^2+\kappa_{2m}\varsigma_{2m}^2+\kappa_{3m}\varsigma_{3m}^2+\kappa_{4m}\varsigma_{1m}\varsigma_{2m}+\kappa_{5m}\varsigma_{1m}\varsigma_{3m}+\kappa_{6m}\varsigma_{2m}\varsigma_{3m}$$

$$+\frac{\chi_{1m}}{s_{44m}}\left[\left(\frac{\partial\varsigma_{1m}}{\partial\varphi}\right)^2+\left(\frac{\partial\varsigma_{1m}}{\partial\nu}\right)^2\right]+\frac{\chi_{2m}}{s_{44m}}\left[\left(\frac{\partial\varsigma_{2m}}{\partial\varphi}\right)^2+\left(\frac{\partial\varsigma_{2m}}{\partial\nu}\right)^2\right]$$

$$+\frac{\chi_{3m}}{s_{44m}}\left[\left(\frac{\partial\varsigma_{3m}}{\partial\varphi}\right)^2+\left(\frac{\partial\varsigma_{3m}}{\partial\nu}\right)^2\right]+\frac{\chi_{4m}}{s_{44m}}\left(\frac{\partial\varsigma_{1m}}{\partial\varphi}\frac{\partial\varsigma_{2m}}{\partial\varphi}+\frac{\partial\varsigma_{1m}}{\partial\nu}\frac{\partial\varsigma_{2m}}{\partial\nu}\right)$$

$$+\frac{\chi_{5m}}{s_{44m}}\left(\frac{\partial\varsigma_{1m}}{\partial\varphi}\frac{\partial\varsigma_{3m}}{\partial\varphi}+\frac{\partial\varsigma_{1m}}{\partial\nu}\frac{\partial\varsigma_{3m}}{\partial\nu}\right)+\frac{\chi_{6m}}{s_{44m}}\left(\frac{\partial\varsigma_{2m}}{\partial\varphi}\frac{\partial\varsigma_{3m}}{\partial\varphi}+\frac{\partial\varsigma_{2m}}{\partial\nu}\frac{\partial\varsigma_{3m}}{\partial\nu}\right),$$

$$W_m=4\int_0^{\pi/2}\int_0^{\pi/2}\left(\Phi_{1m}\varsigma_{1m}^2+\Phi_{2m}\varsigma_{2m}^2+\Phi_{3m}\varsigma_{2m}^2\right)d\varphi\,d\nu$$

$$+4\int_0^{\pi/2}\int_0^{\pi/2}\left(\Phi_{4m}\,\varsigma_{1m}\,\varsigma_{2m}+\Phi_{5m}\,\varsigma_{1m}\,\varsigma_{3m}+\Phi_{6m}\,\varsigma_{2m}\,\varsigma_{3m}\right)d\varphi\,d\nu$$

$$+\frac{4}{s_{44m}}\int_0^{\pi/2}\int_0^{\pi/2}\frac{\Psi_{1m}}{s_{44m}}\left[\left(\frac{\partial\varsigma_{1m}}{\partial\varphi}\right)^2+\left(\frac{\partial\varsigma_{1m}}{\partial\nu}\right)^2\right]d\varphi\,d\nu$$

$$+\frac{4}{s_{44m}}\int_0^{\pi/2}\int_0^{\pi/2}\frac{\Psi_{2m}}{s_{44m}}\left[\left(\frac{\partial\varsigma_{2m}}{\partial\varphi}\right)^2+\left(\frac{\partial\varsigma_{2m}}{\partial\nu}\right)^2\right]d\varphi\,d\nu$$

$$+\frac{4}{s_{44m}}\int_0^{\pi/2}\int_0^{\pi/2}\frac{\Psi_{3m}}{s_{44m}}\left[\left(\frac{\partial\varsigma_{3m}}{\partial\varphi}\right)^2+\left(\frac{\partial\varsigma_{3m}}{\partial\nu}\right)^2\right]d\varphi\,d\nu$$

$$+\frac{4}{s_{44m}}\int_0^{\pi/2}\int_0^{\pi/2}\frac{\Psi_{4m}}{s_{44m}}\left(\frac{\partial\varsigma_{1m}}{\partial\varphi}\frac{\partial\varsigma_{2m}}{\partial\varphi}+\frac{\partial\varsigma_{1m}}{\partial\nu}\frac{\partial\varsigma_{2m}}{\partial\nu}\right)d\varphi\,d\nu$$

$$+\frac{4}{s_{44m}}\int_0^{\pi/2}\int_0^{\pi/2}\frac{\Psi_{5m}}{s_{44m}}\left(\frac{\partial\varsigma_{1m}}{\partial\varphi}\frac{\partial\varsigma_{3m}}{\partial\varphi}+\frac{\partial\varsigma_{1m}}{\partial\nu}\frac{\partial\varsigma_{3m}}{\partial\nu}\right)d\varphi\,d\nu$$

$$+\frac{4}{s_{44m}}\int_0^{\pi/2}\int_0^{\pi/2}\frac{\Psi_{6m}}{s_{44m}}\left(\frac{\partial\varsigma_{2m}}{\partial\varphi}\frac{\partial\varsigma_{3m}}{\partial\varphi}+\frac{\partial\varsigma_{2m}}{\partial v}\frac{\partial\varsigma_{3m}}{\partial v}\right)d\varphi\, dv$$

$$\varsigma_{1m}=-\frac{p_1\,\rho_{2me}\,R_2 r_S^{c_{3m}-1}}{\varsigma_m}\left[1-c_{3m}\left(\frac{1}{2}+\ln r_S\right)\right],$$

$$\varsigma_{2m}=\frac{p_1\,\rho_{2me}\,R_2}{\varsigma_m}\left[\frac{4}{3}-\ln r_S-\left(\frac{4}{3}-\ln r_S\right)\left(\frac{1}{2}+\ln r_S\right)\right],$$

$$\varsigma_{3m}=-\frac{p_1\,\rho_{2me}\,R_2\,r_S^{c_{3m}}}{\varsigma_m}\left[c_{3m}\left(\frac{4}{3}-\ln r_S\right)-\left(\frac{1}{3}-\ln r_S\right)\right],$$

$$\varsigma_m=R_2\,r_S^{c_{3m}-1}\left(\frac{4}{3}-\ln R_2\right)\left[1-c_{3m}\left(\frac{1}{2}+\ln r_S\right)\right]$$

$$-R_2^{c_{3m}}\left[\frac{4}{3}-\ln r_S-\left(\frac{1}{3}-\ln r_S\right)\left(\frac{1}{2}+\ln r_S\right)\right]$$

$$-r_S^{c_{3m}}\left(\frac{1}{2}+\ln R_2\right)\left[c_{3m}\left(\frac{4}{3}-\ln r_S\right)-\left(\frac{1}{3}-\ln r_S\right)\right]. \tag{82}$$

Envelope: The thermal stresses in the envelope for the subscript eB are given by Equations (69), (72), and (73). If $\beta_p \neq \beta_e = \beta_m$, then $|u'_{1e}|$ is required to represent a decreasing function of $r \in \langle R_1, R_2\rangle$. With regard to the dimensions of components of accurate three-component materials, we get $C_{2e} \neq C_{1e} = C_{3e} = 0$ concerning Equation (52). From Equations (18), (31), (31), (52)-(56), we get:

$$\varepsilon'_{11e}=-\frac{p_1\,c_{3e}}{\varsigma_e}\left(\frac{r}{R_1}\right)^{c_{3e}-1},\ \varepsilon'_{22e}=\varepsilon'_{33e}=-\frac{p_1}{\varsigma_e}\left(\frac{r}{R_1}\right)^{c_{3e}-1},$$

$$\varepsilon'_{12e}=s_{44e}\,\sigma'_{12e}=-\frac{1}{\varsigma_e}\left(\frac{r}{R_1}\right)^{c_{3e}-1}\frac{\partial p_1}{\partial\varphi},$$

$$\varepsilon'_{13e}=s_{44e}\,\sigma'_{13e}=-\frac{1}{\varsigma_e}\left(\frac{r}{R_1}\right)^{c_{3e}-1}\frac{\partial p_1}{\partial v},$$

$$\sigma'_{11e} = -p_1 \left(\frac{r}{R_1} \right)^{c_{3e}-1},$$

$$\sigma'_{22e} = \sigma'_{33e} = -\frac{p_1 \left(c_{1e} - c_{2e} c_{3e} \right)}{\varsigma_e} \left(\frac{r}{R_1} \right)^{c_{3e}-1},$$

$$w_e = \left(\frac{1}{\varsigma_e R_1^{c_{3e}-1}} \right)^2 \left\{ \kappa_{2e} p_1^2 + \frac{\chi_{2e}}{s_{44e}} \left[\left(\frac{\partial p_1}{\partial \varphi} \right)^2 + \left(\frac{\partial p_1}{\partial \nu} \right)^2 \right] \right\},$$

$$W_e = \left(\frac{2}{\varsigma_e R_1^{c_{3e}-1}} \right)^2 \int_0^{\pi/2} \int_0^{\pi/2} \left\{ \Phi_{2e} p_1^2 + \frac{\Psi_{2e}}{s_{44e}} \left[\left(\frac{\partial p_1}{\partial \varphi} \right)^2 + \left(\frac{\partial p_1}{\partial \nu} \right)^2 \right] \right\} d\varphi\, d\nu, \quad (83)$$

where s_{44e}, c_{ie} ($i = 1,2,3$), p_1, κ_{2e}, χ_{2e}, Φ_{2e}, Ψ_{2e} are given by Equations (16), (11), (26), (54) and (56), respectively. The coefficient ς_e is derived as:

$$\varsigma_e = \left(c_{3e} + c_{2e} \right) c_{3e} - 2c_{2e}. \quad (84)$$

From $(\varepsilon'_{22e})_{r=R_i} = -p_1 \rho_{ime}$ ($i = 1,2$) and Equation (83), the coefficient ρ_{ime} is derived as:

$$\rho_{1me} = \frac{1}{\varsigma_e},$$

$$\rho_{2me} = \frac{1}{\varsigma_e} \left(\frac{R_2}{R_1} \right)^{c_{3e}-1}. \quad (85)$$

Particle: The thermal stresses in the particle for the subscripts p and pB are given by Equations (75)-(77), whereas p_1 in Equation (75) is provided by Equation (26).

Condition $\beta_p = \beta_e \neq \beta_m$

Matrix: The thermal stresses in the cell matrix for the subscript *m*, *mB* are given by Equations (58), (61)-(63), where p_2 in Equation (58), (61)-(63) is given by Equation (27).

Envelope: The thermal stresses in the envelope for the subscript mB are given by Equations (69)-(73). If $\beta_p = \beta_e \neq \beta_m$, then $|u'_{1e}|$ is required to represent an increasing function of $r \in \langle R_1, R_2 \rangle$, and then we get $C_{1e} \neq C_{2e} = C_{3e} = 0$; $C_{3e} \neq C_{1e} = C_{2e} = 0$ with respect to Equation (52). Consequently, such a combination exhibits a minimum value of W_T (see Equation (42)).

Condition $C_{1e} \neq C_{2e} = C_{3e} = 0$. From Equations (18), (31), (32), (52)-(56), we get:

$$\varepsilon'_{11e} = -\frac{p_2}{\varsigma_e}\left(\frac{1}{3} - \ln r\right), \varepsilon'_{22e} = \varepsilon'_{33e} = -\frac{p_2}{\varsigma_e}\left(\frac{4}{3} - \ln r\right),$$

$$\varepsilon'_{12e} = s_{44e}\,\sigma'_{12e} = -\frac{1}{\varsigma_e}\left(\frac{4}{3} - \ln r\right)\frac{\partial p_2}{\partial \varphi},$$

$$\varepsilon'_{13e} = s_{44e}\,\sigma'_{13e} = -\frac{1}{\varsigma_e}\left(\frac{4}{3} - \ln r\right)\frac{\partial p_2}{\partial \nu},$$

$$\sigma'_{11e} = -\frac{p_2}{\varsigma_e}\left[\frac{c_{1e} - 7c_{2e}}{3} - (c_{1e} - c_{2e})\ln r\right],$$

$$\sigma'_{22e} = \sigma'_{33e} = -\frac{p_2}{\varsigma_e}\left[\frac{4c_{1e} - c_{2e}}{3} - (c_{1e} - c_{2e})\ln r\right],$$

$$w_e = \frac{1}{\varsigma_e^2}\left\{\kappa_{1e}\,p_2^2 + \frac{\chi_{1e}}{s_{44e}}\left[\left(\frac{\partial p_2}{\partial \varphi}\right)^2 + \left(\frac{\partial p_2}{\partial \nu}\right)^2\right]\right\},$$

$$W_e = \left(\frac{2}{\varsigma_e}\right)^2 \int_0^{\pi/2}\int_0^{\pi/2}\left\{\Phi_{1e}\,p_2^2 + \frac{\Psi_{1e}}{s_{44e}}\left[\left(\frac{\partial p_2}{\partial \varphi}\right)^2 + \left(\frac{\partial p_2}{\partial \nu}\right)^2\right]\right\} d\varphi\, d\nu, \quad (86)$$

where s_{44e}, c_{ie} $(i = 1,2,3)$, p_2, κ_{1e}, χ_{1e}, Φ_{1e}, Ψ_{1e} are given by Equations (16), (11), (27), (54) and (56), respectively. The coefficient ς_e is derived as:

$$\varsigma_e = \frac{c_{1e} - 7c_{2e}}{3} - (c_{1e} - c_{2e})\ln R_2. \quad (87)$$

About $(\varepsilon'_{22e})_{r=Ri} = - p_2\,\rho_{ipe}$ $(i = 1,2)$ and Equation (86), the coefficient ρ_{ipe} is derived as:

$$\rho_{ipe} = \frac{1}{\varsigma_e}\left(\frac{4}{3} - \ln R_i\right), i = 1,2. \tag{88}$$

Condition $C_{3e} \neq C_{1e} = C_{2e} = 0$. With regard to Equations (18), (31), (32), (52)-(56), we get:

$$\varepsilon_{11e}^{'} = -\frac{p_2}{\varsigma_{eB}\, r}, \varepsilon_{22e}^{'} = \varepsilon_{33e}^{'} = -\frac{p_2}{\varsigma_{eB}\, r}\left(\frac{1}{2} + \ln r\right),$$

$$\varepsilon_{12e}^{'} = s_{44e}\,\sigma_{12e}^{'} = -\frac{1}{\varsigma_e\, r}\left(\frac{1}{2} + \ln r\right)\frac{\partial p_2}{\partial \varphi},$$

$$\varepsilon_{13eB}^{'} = s_{44e}\,\sigma_{13eB}^{'} = -\frac{1}{\varsigma_e\, r}\left(\frac{1}{2} + \ln r\right)\frac{\partial p_2}{\partial v},$$

$$\sigma_{11e}^{'} = -\frac{p_2}{\varsigma_e\, r}\left(c_{1e} - 2c_{2e}\ln r\right),$$

$$\sigma_{22e}^{'} = \sigma_{33e}^{'} = -\frac{p_2}{\varsigma_e\, r}\left(\frac{c_{1e} - 2c_{2e}}{2} + c_{1e}\ln r\right),$$

$$w_e = \frac{1}{\varsigma_e^2}\left\{\kappa_{3e}\, p_2^2 + \frac{\chi_{3e}}{s_{44e}}\left[\left(\frac{\partial p_2}{\partial \varphi}\right)^2 + \left(\frac{\partial p_2}{\partial v}\right)^2\right]\right\},$$

$$W_e = \left(\frac{2}{\varsigma_e}\right)^2 \int_0^{\pi/2}\int_0^{\pi/2}\left\{\Phi_{3e}\, p_2^2 + \frac{\Psi_{3e}}{s_{44e}}\left[\left(\frac{\partial p_2}{\partial \varphi}\right)^2 + \left(\frac{\partial p_2}{\partial v}\right)^2\right]\right\} d\varphi\, dv,$$

$$\varsigma_e = \frac{c_{1e} - 2c_{2e}\ln R_2}{R_2}, \rho_{ipe} = \frac{1}{\varsigma_e R_i}\left(\frac{1}{2} + \ln R_i\right), i = 1,2. \tag{89}$$

Particle: The thermal stresses in the particle for the subscript *pB* are given by Equations (75)-(77). From Equations (3)-(6), (7)-(10), (18), (19) and (30), we get:

$$\varepsilon_{11p}^{'} = -\, p_2\, \rho_{1pe}\, \lambda_{1p}\left(\frac{r}{R_1}\right)^{\lambda_{1p}-1},$$

$$\varepsilon_{22p}^{'} = \varepsilon_{33p}^{'} = -\, p_2\, \rho_{1pe}\left(\frac{r}{R_1}\right)^{\lambda_{1p}-1},$$

$$\varepsilon'_{12p} = s_{44p}\sigma'_{12p} = -\rho_{1pe}\left(\frac{r}{R_1}\right)^{\lambda_{1p}-1}\frac{\partial p_2}{\partial \varphi},$$

$$\varepsilon'_{13p} = s_{44p}\sigma'_{13p} = -\rho_{1pe}\left(\frac{r}{R_1}\right)^{\lambda_{1p}-1}\frac{\partial p_2}{\partial v},$$

$$\sigma'_{11p} = -p_2\,\rho_{1pe}\,\xi_{3p}\left(\frac{r}{R_1}\right)^{\lambda_{1p}-1},$$

$$\sigma'_{22p} = \sigma'_{33p} = -p_2\,\rho_{1pe}\,\xi_{5p}\left(\frac{r}{R_1}\right)^{\lambda_{1p}-1},$$

$$w_p = \kappa_p\left(\frac{r}{R_1}\right)^{2(\lambda_{1p}-1)}, \; W_p = \frac{4R_1^3}{2\lambda_{1p}+1}\int_0^{\pi/2}\int_0^{\pi/2}\kappa_p\, d\varphi\, dv, \tag{90}$$

where p_2, z_{ip} $(i = 3,5)$, ρ_{1pe} are given by Equations (27); (77); (88), and (89), respectively, and κ_p is derived as:

$$\kappa_p = \rho_{1pe}^2\left\{\frac{E_p}{(1+\mu_p)(1-2\mu_p)}\left[\frac{\lambda_{1p}^2(1-\mu_p)}{2} + 2\lambda_{1p}\mu_p + 1\right]\right.$$

$$\left. + \frac{1}{s_{44p}}\left[\left(\frac{\partial p_2}{\partial \varphi}\right)^2 + \left(\frac{\partial p_2}{\partial v}\right)^2\right]\right\}. \tag{91}$$

Mathematical Model of Cracking

Figures 5a and 5b show a solid continuum with the volume V in the Cartesian system $(Ox_1x_2x_3)$ without and with a crack, respectively. The crack is formed in the plane x_1x_2. The shaded area in Figure 5a represents cuts of the solid continuum in the planes x_1x_2, x_1x_3, and x_2x_3, where $x \subset x_1x_2$, $\varphi = \angle\,(x_1, x_2) \in \langle 0, 2\pi\rangle$. The curves 1, 2, 3, and 4 outline the cuts in the planes x_1x_3, x_1x_2, x_2x_3, and xx_3, respectively. Additionally, P and P'' represent points on the axis x_1 and on the curve 4, respectively.

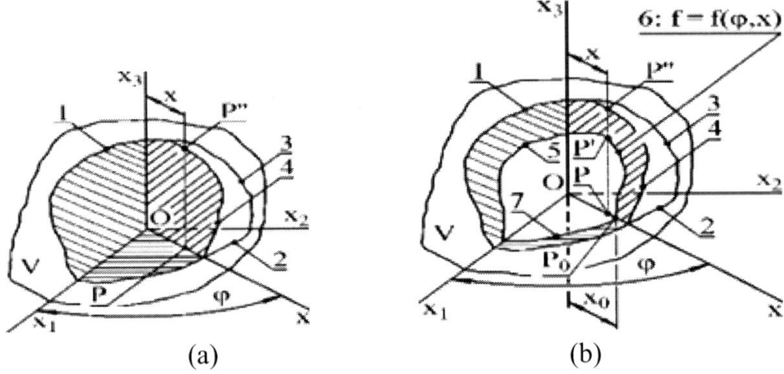

Figure 5. The solid continuum has a general shape with volume V in the Cartesian system ($Ox_1x_2x_3$): (a) without and (b) with a crack. The crack is formed in the plane x_1x_2. The shaded area represents cuts of the solid continuum in the planes x_1x_3, x_1x_2, x_2x_3, and xx_3, where $x \subset x_1x_2$, $\varphi = \angle\,(x_1, x_2) \in \langle 0, 2\pi \rangle$.The curves *1*, *2*, *3*, and 4 outline the cuts in the planes x_1x_3, x_1x_2, x_2x_3, and xx_3, respectively. Points P and P'' represent points on the axis x_1 and on curve *4*, respectively. Figures *5* and *6* represent the crack shape in the planes x_1x_3 and xx_3 (see Figure 3). The crack shape in xx_3 is defined by the function $f = f(\varphi, x, x_3)$ of the cylindrical coordinates (φ, x, x_3). Curve *7* defines the position of the crack tip in the x_1x_2 plane. Consequently, the point P_0 represents the crack tip in the plane x_1x_2.

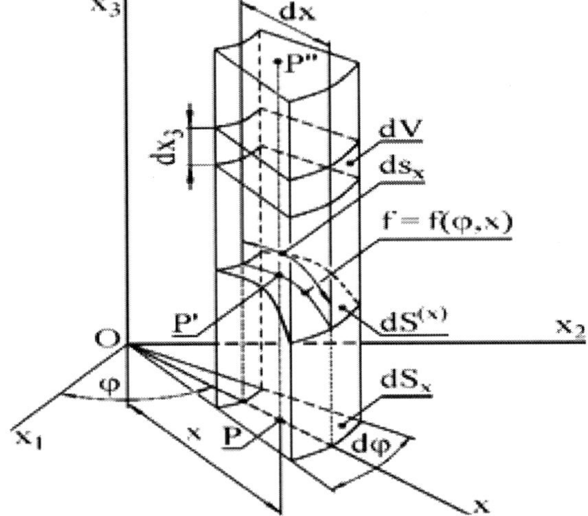

Figure 6. The infinitesimal prism with height $|PP''|$ (see Figure 5b) and the small surface element dSx $= x\, d\varphi dx$ in the x_1x_2 plane. The small crack surface $dS^{(x)}$ is related to the infinitesimal crack length ds_x at point P' (see Figure 5b). The function $f = f(\varphi, x)$ defines the crack shape in the x_1x_2 plane.

Figures 5 and 6 represent the crack shape in x_1x_3 and xx_3 planes (see Figure 3). The crack shape in the xx_3 plane is defined by the function $f=f(\varphi, x, x_3)$ in cylindrical coordinates (φ,x,x_3). Curve 7 illustrates the position of the crack tip in the x_1x_2 plane. Consequently, the point P_0 represents the crack tip in the x_1x_2 plane. The thermal stresses are derived using spherical coordinates (r,φ,ν) (see Figure 2). Therefore, the following transformations need to be considered:

$$r=\sqrt{x^2+x_3^2}\ ,\nu=\arctan\left(\frac{x}{x_3}\right). \tag{92}$$

Let the thermal-stress-induced elastic energy density $w = w(r,\varphi,\nu)$ be determined in cylindrical coordinates (φ, x, x_3), so we have $w = w(\varphi, x, x_3)$. The elastic energy $W=\int w dV$ accumulated in the volume V (see Figure 5a) tends to be released by crack formation in the x_1x_2 plane. The same principle applies to the differential energy $dW = dW(w)$, accumulated in the volume $dV=\int_P^{P''} dS_x\, dx_3$ of the infinitesimal prism in Figure 6, where $dS_x = x\, d\varphi\, dx$.

Consequently, we get:

$$dW=\int_P^{P''} w dV=\int_P^{P''} w dS_x\, dx_3 = W_c\, x\, d\varphi\, dx\ ,\ \ W_{CI}=\int_P^{P''} w\, dx_3 \tag{93}$$

The crack is formed in the plane x_1x_2 provided that the condition:

$$\int_P^{P''} w(x,\varphi,x_3)dx_3=\int_P^{P'''} w(x,\varphi,-x_3)dx_3,$$
$$\varphi\in\langle 0,2\pi\rangle \tag{94}$$

It is valid for $\varphi \in\langle 0,2\pi\rangle$, where P''' is the intersection point of the line PP''' with the solid continuum surface for $x_3 < 0$. Accordingly, x_1x_2 is required to represent a symmetry plane with respect to the curve integral $W_c = W_c(\varphi,x,x_3)$ for $x_3 >0$ and $x_3 < 0$.

The energy dW is in equilibrium with the energy $dW_x = \gamma dS^{(x)}$, which creates the infinitesimal crack surface $dS^{(x)} = ds_x\, x\, d\varphi$ (see Figure 6). The surface energy density γ and the infinitesimal crack length ds_x are derived as:

$$\gamma = \frac{K_{IC}^2}{E}, \tag{95}$$

$$ds_x = dx\sqrt{1+\left(\frac{\partial f}{\partial x}\right)^2}, \tag{96}$$

where K_{IC} is the fracture toughness. From the equilibrium condition $dW = dW_x$ (see Equations (95), (96)), we get:

$$\frac{\partial f}{\partial x} = \pm\frac{E}{K_{IC}^2}\sqrt{W_{CI}^2-\left(\frac{K_{IC}^2}{E}\right)^2}. \tag{97}$$

The thermal stresses, the elastic energy density w, and the curve integral W_{CI} represent decreasing functions of the variable x. Consequently, $f = f(x,\varphi)$ is a decreasing function of x. Accordingly, the sign '-' in Equation (97) is considered. The formula (97) is also valid for $W_{CI} - K_{IC}^2/E \geq 0$. The model system in Figure 1 is symmetric. Therefore, the thermal stresses need only be investigated within one-eighth of the cubic cell (see Figure 7), i.e., for $\varphi\in\langle 0,\pi/2\rangle$, $\nu\in\langle 0,\pi/2\rangle$. The crack formation in this chapter is determined in the cell matrix, i.e., under the condition $\beta_e < \beta_m$. In the case of the cell matrix, the function $f_m = f_m(x,\varphi)$ (i.e., the curve 6 in Figure 5a) of the variable $x \in\langle R_2, x_0\rangle$ for $R_2 > R_{2c}$ (see Figure 8a) is derived as:

$$f_m = \frac{E_m}{K_{ICm}^2}\left[C_m - \int\sqrt{W_{C\,\mathrm{Im}}^2-\left(\frac{K_{ICm}^2}{E_m}\right)^2}\,dx\right],\ x\in\langle R_2, x_0\rangle,\ R_2 > R_{2c}, \tag{98}$$

where Young's modulus E_m and the fracture toughness K_{ICm} are related to the matrix, R_{2c} is the critical envelope radius, and C_m is an integration constant.

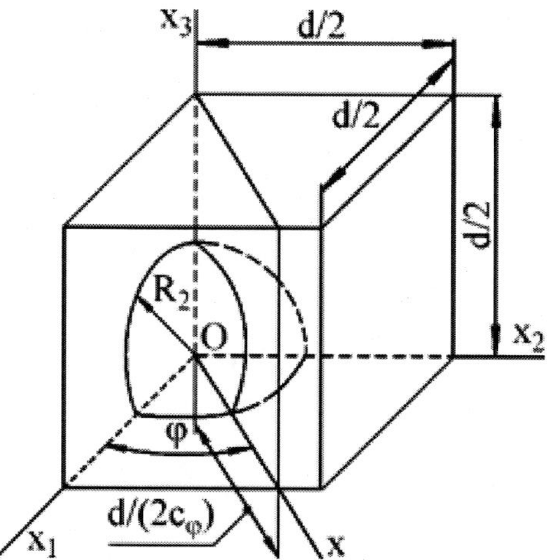

Figure 7. One-eighth of the cubic cell, with the central spherical envelope having a R_2, where d is the cubic cell dimension. The general position of the axis x in the plane x_1x_2 is determined by the angle $\varphi\in\langle 0,\pi/2\rangle$. The coefficient $c_\varphi = c_\varphi\,(\iota)$ is given by Equation (2).

The curve integral $W_{C\mathrm{Im}} = W_{C\mathrm{Im}}(\varphi,x)$ in the cell matrix along the abscissa PP'' (see Figure 8b) is derived as:

$$W_{C\mathrm{Im}} = \int_{P}^{P''} \left(w_m + w_{mB}\right) dx_3 = \int_{0}^{d/2} \left(w_m + w_{mB}\right) dx_3 \;, x \in \left\langle R_2, \frac{d}{2c_\varphi} \right\rangle. \quad (99)$$

The critical envelope radius $R_{2c} = R_{2c}(\varphi)$ is determined by the following condition:

$$\left[W_{C\mathrm{Im}}\left(x,\varphi,R_2\right)\right]_{x=R2} - \frac{K_{ICm}^2}{E_m} = 0, \quad (100)$$

which represents an equation with the variable R_2 and the parameter φ, where $R_{2c} = R_{2c}(\varphi)$ represents a root of Equation (100). If $R_2 = R_{2c}$, an infinitesimal crack with the length dx along the x-axis is formed in the plane x_1x_2. Accordingly, $R_2 = R_{2c}(\varphi)$ defines a limit state for crack formation.

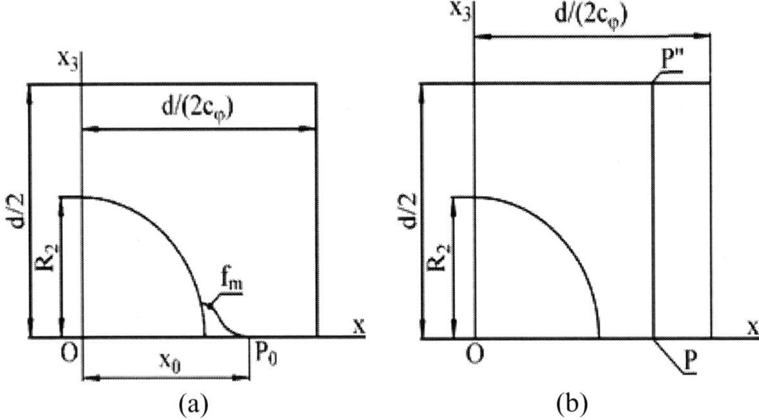

Figure 8. (a) The crack shape in the plane x_1x_2 (see Figure 7) with the crack tip P_0. (see Figure 5b). (b) The abscissa PP'' (see Figure 6) for a definition of the curve integral W_{CIm} for $x\in\langle R_2, d / c_\varphi\rangle$ (see Equation (99)), where the coefficient $c_\varphi = c_\varphi(\varphi)$ is given by Equation (2).

This limit state corresponds to a single value of the variable $\varphi\in\langle 0,\pi/2\rangle$. The maximum value R_{2cmax} of the function $R_{2c}= R_{2c}(\varphi)$ defines the limit state corresponding to each value of $\varphi\in\langle 0,\pi/2\rangle$. If $R_2 > R_{2c}(\varphi)$, then $x_0 = x_0(\varphi,R_2)$ is determined by the following condition

$$W_{C\mathrm{Im}}\left(x,\varphi,R_2\right)-\frac{K_{ICm}^2}{E_m}=0,\quad R_2>R_{2c} \tag{101}$$

which represents an equation with the variable x and the parameters φ, R_2. The root $x_0 = x_0(\varphi, R_2)$ of Equation (100) defines the shape of the crack in the crack formation plane, i.e., the curve *7* in Figure 5b. Finally, the integration constant $C_m = C_m(\varphi,R_2)$ for the parameters $\varphi\in\langle 0,\pi/2\rangle$, $R_2 > R_{2c}(\varphi)$, which is determined by the boundary condition $[f_m(x,\varphi)]_{x=x0} = 0$, is derived as:

$$C_m=\left[\int\sqrt{W_{C\mathrm{Im}}^2-\left(\frac{K_{ICm}^2}{E_m}\right)^2}\,dx\right]_{x=x0}. \tag{102}$$

On the one hand, the thermal stresses, along with W_{CIm} (see Equation (99)), are modified (changed) during the crack formation for $R_2 > R_{2c}$. If analytical or computational methods determine such a modification, this determination can be applied to Equations (100)-(102). On the other hand, the curve integral W_{CIm} (see Equation (99)), which does not account for this modification, is applied in the case of a high-speed crack formation in brittle ceramic materials.

Conclusion

This chapter presents a mathematical model of thermal stress interactions and crack formation in a material system. Thermal stresses, caused by cooling and differing expansion coefficients, interact within a system of spherical particles and envelopes in an isotropic matrix. The model, based on continuum mechanics and the superposition method, incorporates microstructural parameters. A crack model compares thermal stress energy with the energy needed to form a crack, determining the critical envelope radius for matrix crack formation. The results apply to research and engineering in material stress modeling and energy interactions.

Disclaimer

None.

References

[1] Ceniga L. (2007). *Analytical Models of Thermal Stresses in Composite Materials II*, Nova Science Publishers.

[2] Mura T. (1987). *Micromechanics of Defects in Solids*, Martinus Nijhoff Publishers.

[3] Ceniga L. (2022). *Mathematical Determination of Residual Stresses in Two-Component Materials*, Nova Science Publishers.

[4] Ceniga, L.(2024a). Mathematical Model of Thermal Stress Induced Strengthening in Porous Composites. *New Research on Thermal Stresses*, Nova Science Publishers, 1-43.

[5] Brdicka M., Samek L. & Sopko B. (2000). Mechanics of Continuum, Academia.

[6] Ceniga L. (2024b). *Thermal-Stress-Field Interactions in Composite Materials I*, Superposition Method, Lambert Academic Publishing.

[7] Ceniga L. (2012). *Analytical Models of Thermal Stresses in Composite Materials III*, Nova Science Publishers.

[8] Skocovsky P., Bokuvka O. & Palcek P. (1996). *Materials Science*, EDIS, TU Zilina.

[9] Rektorys K. (1973). *Review of Applied Mathematics*, SNTL, Prague.

Chapter 3

Three-Dimensional Analysis of the Thermally Stressed State of Layered Cylindrical Shells

Alexandr V. Marchuk*, **DSc**

Department of Strength of Materials and Engineering Science,
National Transport University, Kiev, Ukraine

Abstract

The chapter explores the three-dimensional analysis of layered composite cylindrical shells under complex thermal conditions. It introduces new methods based on Reissner's variational principle and a semi analytical finite element method to address challenges such as localized heating, delamination, and intricate boundary conditions. These advancements provide accurate solutions, overcoming the limitations of traditional two-dimensional models.

Keywords: cylindrical shells, thermally stressed state, analytical solution, semi analytical solution, internal heat source

Introduction

The calculation of layered shells under thermal effects has been the subject of extensive research, as evidenced by numerous publications in this field. A review of this research can be found, in particular, in the works of Ootao et al. [1], Brischetto & Carrera [2], Tokovyy et al. [3], and Musii et al. [4]. Shear

---

* Corresponding Author's Email: mksmntu386@gmail.com

In: Thermal Modeling Reimagined
Editors: Pankaj Thakur and Jatinder Kaur
ISBN: 979-8-89530-463-1

models are primarily used for the study, as exemplified by Ootao et al. [1]. Some models also utilize a linear approximation of displacements and temperature distribution along the shell thickness, as shown in the work of Musii et al. [4]. In the exact three-dimensional formulation, the problem is solved in Tokovyy et al. [3], where the axisymmetric bending of a long shell is considered. The need for an exact solution to the heat conduction equations is highlighted in Brischetto & Carrera [2]. In Grigorenko et al. [5, 6], the thermal stress state of cylindrical shells is studied using a high-precision numerical approach. To the authors' knowledge, no previous work has been dedicated to three-dimensional analytical calculations of temperature effects in finite-sized, layered orthotropic shells. Moreover, the semi analytical finite element method has yet to be extended to encompass the analysis of the thermally stressed state of layered shells, where the sought functions are determined through the solution of the corresponding systems of differential equations.

Problem Statement

The primary objective of this research is to develop a comprehensive three-dimensional analytical method for analyzing the thermo-elastic state of hinged-supported, layered composite cylindrical shells. Using this approach, three-dimensional analytical solutions will be derived for the first time, addressing scenarios of thermal loading, heat flow, and internal heat sources, marking a significant milestone in analyzing these complex structures. A new version of the semi analytical finite element method is also being developed, in which linear polynomials approximate the required functions in the shell plane, and their thickness distribution is determined based on the analytical solution of the corresponding system of differential equations. Based on this approach, problems such as local temperature loading, delamination with separation along part of the surface, and complex boundary conditions on the shell's outer surface will be considered. The solutions are based on dividing a cylindrical shell along its thickness by concentric surfaces into several constituent cylindrical shells (Figure 1), each thin enough to neglect the change in curvature over the thickness.

Thus, the thermally stressed state of the initial shell is described with discrete consideration of the change in curvature over the thickness. This makes it possible to extend the approaches for solving problems related to the thermal stress state of plates and shallow shells (Marchuk et al. [7, 8]) to calculating cylindrical shells. The analysis of the resulting solutions will reveal that the loss of accuracy in this scenario is not significant. Similarly, the dynamics problems were considered in Marchuk et al. [9]. Using Reissner's variational principle, we obtain a system of integro-differential equilibrium equations without introducing any approximation, similar to the method used by Marchuk & Gnedash [10]. For the particular case of hinge-supported cylindrical shells with thermal load distributed according to the trigonometric law, the system of integro-differential equations of equilibrium admits an analytical solution. The sought functions in the shell's plane are approximated in more intricate loading cases and boundary conditions. This leads to a system of differential equations, which is then analytically solved. Similarly, statics problems were considered in the works of Marchuk et al. [11]. To test the proposed method, a method with a polynomial approximation of displacements along the thickness of the shell is used.

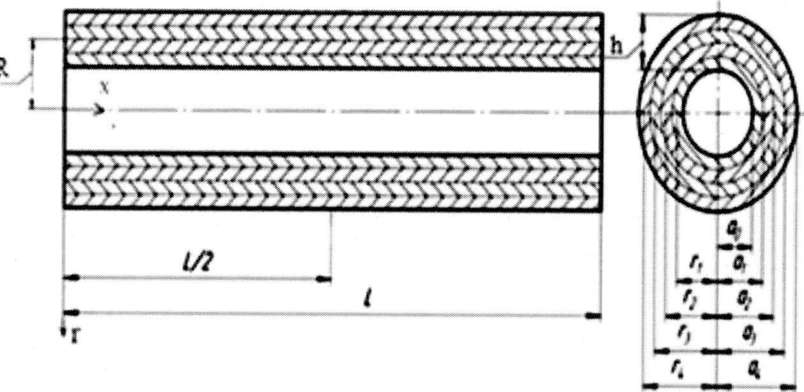

Figure 1. Scheme of the shell.

Considering a cylindrical coordinate system, the axes (with lower index 1) are directed within the plate plane, and the axis (with lower index 3) is directed downward along the thickness. A comma at the index level indicates the operation of differentiation.

Calculation Method

Let us write the displacement vector and transverse components of the stress tensor, as well as the temperature distribution, in the following form:

$$U_x^{(k)}(x,r)=V_1^{(k)}(x)f_1^{(k)}(r)\,,\ U_r^{(k)}(x,r)=V_3^{(k)}(x)f_3^{(k)}(r)$$

$$\sigma_{xr}^{(k)}(x,r)=V_4^{(k)}(x)f_4^{(k)}(r)\ \sigma_{rr}^{(k)}(x,r)=V_6^{(k)}(x)f_6^{(k)}(r)\,,$$

$$T^{(k)}(x,r)=V_7^{(7)}(x)f_7^{(k)}(r) \tag{1}$$

Using Cauchy relations and Equation (1), expressions for strains are written as:

$$e_{xx}^{(k)}=V_{1,1}^{(k)}f_1^{(k)}\,,\ e_{\theta\theta}^{(k)}=\frac{1}{r}V_3^{(k)}f_3^{(k)}\,,\ e_{rr}^{(k)}=V_3^{(k)}f_{3,3}^{(k)}\,,$$

$$2e_{xr}^{(k)}=V_1^{(k)}f_{1,3}^{(k)}+V_{3,1}^{(k)}f_3^{(k)} \tag{2}$$

Hooke's law and Equation (2) allow one to determine the stresses $\sigma_{xx}^{(k)}$ and $\sigma_{\theta\theta}^{(k)}$ with account of thermal loading:

$$\sigma_{xx}^{(k)}=B_{11}^{(k)}V_{1,1}^{(k)}f_1^{(k)}+B_{12}^{(k)}\frac{1}{r}V_3^{(k)}f_3^{(k)}+B_{13}^{(k)}V_6^{(k)}f_6^{(k)}-(\alpha_1 B_{11}^{(k)}+\alpha_2 B_{12}^{(k)})V_7^{(k)}f_7^{(k)}$$

$$\sigma_{\theta\theta}^{(k)}=B_{21}^{(k)}V_{1,1}^{(k)}f_1^{(k)}+B_{22}^{(k)}\frac{1}{r}V_3^{(k)}f_3^{(k)}+B_{23}^{(k)}V_6^{(k)}f_6^{(k)}-(\alpha_1 B_{21}^{(k)}+\alpha_2 B_{22}^{(k)})V_7^{(k)}f_7^{(k)} \tag{3}$$

The resolving system of equations and the corresponding boundary conditions are found using the Reissner variational principle:

$$\int_L[-V_1^{(k)}V_4^{(k)}f_{1,r}^{(k)}-V_{3,1}^{(k)}V_4^{(k)}f_3^{(k)}+\frac{1}{G_{13}^{(k)}}V_4^{(k)}V_4^{(k)}f_4^{(k)}]dx=0\,,$$

$$\int_L[-B_{13}^{(k)}V_{1,1}^{(k)}V_6^{(k)}f_1^{(k)}-B_{23}^{(k)}\frac{1}{r}V_3^{(k)}V_6^{(k)}f_3^{(k)}-V_3^{(k)}V_6^{(k)}f_{3,3}^{(k)}+B_{33}^{(k)}V_6^{(k)}V_6^{(k)}f_6^{(k)}+$$
$$+(B_{13}^{(k)}\alpha_1^{(k)}+B_{23}^{(k)}\alpha_2^{(k)}+\alpha_3^{(k)})V_7^{(k)}V_6^{(k)}f_7^{(k)}]dx=0\,,$$

$$\int_L[B_{11}^{(k)}V_{1,1}^{(k)}V_{1,1}^{(k)}f_1^{(k)}+B_{12}^{(k)}\frac{1}{r}V_3^{(k)}V_{1,1}^{(k)}f_3^{(k)}+B_{13}^{(k)}V_6^{(k)}V_{1,1}^{(k)}f_6^{(k)}-(\alpha_1 B_{11}^{(k)}+\alpha_2 B_{12}^{(k)})V_7^{(k)}V_{1,1}^{(k)}f_7^{(k)}(z)+$$
$$-\tau_{13}^{(k)}V_1^{(k)}f_{4,3}^{(k)}-\frac{1}{r}\tau_{13}^{(k)}V_1^{(k)}f_4^{(k)}]dx=0\,,$$

$$+V_4^{(k)}V_{3,1}^{(k)}f_4^{(k)} - V_6^{(k)}V_3^{(k)}f_{3,3}^{(k)} - \frac{1}{r}V_6^{(k)}V_3^{(k)}f_3^{(k)}\,]dx = 0 \tag{4}$$

These equations have to be supplemented by the known equations of stationary heat conductivity:

$$\lambda_{xx}^{(k)}T_{,11}^{(k)} + \lambda_{rr}^{(k)}T_{,33}^{(k)} + \lambda_{rr}^{(k)}\frac{1}{r}T_{,3}^{(k)} + q_t^{(k)} = 0 \tag{5}$$

$$\int_L [\frac{1}{r}B_{21}^{(k)}V_{1,1}^{(k)}V_3^{(k)}f_1^{(k)} + B_{22}^{(k)}\frac{1}{r}\frac{1}{r}V_3^{(k)}V_3^{(k)}f_3^{(k)} + \frac{1}{r}B_{23}^{(k)}V_6^{(k)}V_3^{(k)}f_6^{(k)} - \frac{1}{r}(\alpha_1 B_{21}^{(k)} + \alpha_2 B_{22}^{(k)})V_7^{(k)}V_3^{(k)}f_7^{(k)}(z) +$$

This Equation (5) can also be represented in the following form:

$$-\lambda_{rr}^{(k)}T_{,3}^{(k)} - Q_3^{(k)} = 0,$$

$$\lambda_{xx}^{(k)}T_{,11}^{(k)} - Q_{,3}^{(k)} - \frac{1}{r}Q_3^{(k)} + q_t^{(k)} = 0 \tag{6}$$

Taking into account the separation of variables $Q_3^{(k)} = V_8^{(k)}f_8^{(k)}$ $q_t^{(k)} = V_9^{(k)}q^{(k)}$, Equations (6) are transformed as follows:

$$-\lambda_{rr}^{(k)}V_7^{(7)}f_{7,3}^{(7)} - V_8^{(k)}f_8^{(k)} = 0,$$

$$\lambda_{xx}^{(k)}V_{7,11}^{(k)}f_7^{(k)} - V_8^{(k)}f_{8,3}^{(k)} - \frac{1}{r}V_8^{(k)}f_8^{(k)} + V_9^{(k)}q^{(k)} = 0 \tag{7}$$

For hinged support, the required functions along the x-axis can be expressed in the form:

$$V_1^{(k)} = V_4^{(k)} = \cos\frac{\pi mx}{l},\ V_3^{(k)} = V_6^{(k)} = V_7^{(k)} = V_8^{(k)} = V_9^{(k)} = \sin\frac{\pi mx}{l} \tag{8}$$

Equations (4) and (7) can be expressed as follows:

$$f_{1,3}^{(k)} = -f_3^{(k)}\left(\frac{\pi m}{l}\right) + f_4^{(k)}/G_{13}^{(k)},$$

$$f_{3,3}^{(k)} = f_1^{(k)}B_{13}^{(k)}\left(\frac{\pi m}{l}\right) - f_3^{(k)}\frac{1}{r^{(k)}}B_{23}^{(k)} + f_6^{(k)}B_{33}^{(k)} + (B_{13}^{(k)}\alpha_1^{(k)} + B_{23}^{(k)}\alpha_2^{(k)} + \alpha_3^{(k)})f_7^{(k)}$$

$$f_{4,3}^{(k)}=\left[\begin{array}{l} f_1^{(k)}B_{11}^{(k)}\left(\frac{\pi m}{l}\right)^2-f_3^{(k)}\frac{1}{r^{(k)}}B_{12}^{(k)}\left(\frac{\pi m}{l}\right)-f_4^{(k)}\frac{1}{r^{(k)}} \\ -f_6^{(k)}B_{13}^{(k)}+f_7^{(k)}(z)(\alpha_1 B_{11}^{(k)}+\alpha_2 B_{12}^{(k)})\left(\frac{\pi m}{l}\right) \end{array}\right]$$

$$f_{6,3}^{(k)}=-f_1^{(k)}\frac{1}{r^{(k)}}B_{12}^{(k)}\left(\frac{\pi m}{a}\right)+f_3^{(k)}\frac{1}{r^{(k)2}}B_{22}^{(k)}+f_4^{(k)}\left(\frac{\pi m}{a}\right)+f_6^{(k)}\left(\frac{1}{r^{(k)}}B_{23}^{(k)}-\frac{1}{r^{(k)}}\right)-$$

$$-f_7^{(k)}\frac{1}{r^{(k)}}(\alpha_1 B_{21}^{(k)}+\alpha_2 B_{22}^{(k)}),$$

$$f_{7,3}^{(k)}=-f_8^{(k)}/\lambda_{rr}^{(k)},$$

$$f_{8,3}^{(k)}=-\lambda_{xx}^{(k)}\left(\frac{\pi m}{l}\right)^2 f_7^{(k)}-\frac{1}{r^{(k)}}f_8^{(k)}+q^{(k)} \tag{9}$$

A partial solution to the steady-state heat conduction equation has the form:

$$\bar{f}_7^{(k)}=\frac{q}{\lambda_{xx}^{(k)}(\frac{\pi m}{l})^2},\quad \bar{f}_8^{(k)}=0 \tag{10}$$

We obtain a partial solution to the system of differential Equations (9) from the solution of the following system of algebraic equations:

$$0=-\bar{f}_3^{(k)}\left(\frac{\pi m}{l}\right)+\bar{f}_4^{(k)}/G_{13}^{(k)},$$

$$-(B_{13}^{(k)}\alpha_1^{(k)}+B_{23}^{(k)}\alpha_2^{(k)}+\alpha_3^{(k)})\bar{f}_7^{(k)}=\bar{f}_1^{(k)}B_{13}^{(k)}\left(\frac{\pi m}{l}\right)-\bar{f}_3^{(k)}\frac{1}{r^{(k)}}B_{23}^{(k)}+\bar{f}_6^{(k)}B_{33}^{(k)}$$

$$-\bar{f}_7^{(k)}(z)(\alpha_1 B_{11}^{(k)}+\alpha_2 B_{12}^{(k)})\left(\frac{\pi m}{l}\right)=\bar{f}_1^{(k)}B_{11}^{(k)}\left(\frac{\pi m}{l}\right)^2-\bar{f}_3^{(k)}\frac{1}{r^{(k)}}B_{12}^{(k)}\left(\frac{\pi m}{l}\right)-\bar{f}_4^{(k)}\frac{1}{r^{(k)}}-\bar{f}_6^{(k)}B_{13}^{(k)}$$

$$\bar{f}_7^{(k)}\frac{1}{r^{(k)}}(\alpha_1 B_{21}^{(k)}+\alpha_2 B_{22}^{(k)})=-\bar{f}_1^{(k)}\frac{1}{r^{(k)}}B_{12}^{(k)}\left(\frac{\pi m}{a}\right)+\bar{f}_3^{(k)}\frac{1}{r^{(k)2}}B_{22}^{(k)}+\bar{f}_4^{(k)}\left(\frac{\pi m}{a}\right)+\bar{f}_6^{(k)}\left(\frac{1}{r^{(k)}}B_{23}^{(k)}-\frac{1}{r^{(k)}}\right)=0 \tag{11}$$

Now, the solution of the system of differential Equations (9) can be written as follows:

$$f_1^{(k)}=\mu_{11}^{(k)}C_1^{(k)}e^{\beta_1^{(k)}r}+\mu_{13}^{(k)}C_3^{(k)}e^{\beta_3^{(k)}r}+$$

$$+\mu_{14}^{(k)}C_4^{(k)}e^{\beta_4^{(k)}r}+\mu_{16}^{(k)}C_6^{(k)}e^{\beta_6^{(k)}r}+C_7^{(k)}\mu_{17}^{(k)}e^{\beta_7^{(k)}r}+C_8^{(k)}\mu_{18}^{(k)}e^{\beta_8^{(k)}r}+\bar{f}_1^{(k)}$$

$$f_3^{(k)}=\mu_{31}^{(k)}C_1^{(k)}e^{\beta_1^{(k)}r}+\mu_{33}^{(k)}C_3^{(k)}e^{\beta_3^{(k)}r}+$$

$$+\mu_{34}^{(k)}C_4^{(k)}e^{\beta_4^{(k)}r}+\mu_{36}^{(k)}C_6^{(k)}e^{\beta_6^{(k)}r}+C_7^{(k)}\mu_{37}^{(k)}e^{\beta_7^{(k)}r}+C_8^{(k)}\mu_{38}^{(k)}e^{\beta_8^{(k)}r}+\bar{f}_3^{(k)}$$

$$f_4^{(k)}=\mu_{41}^{(k)}C_1^{(k)}e^{\beta_1^{(k)}r}+\mu_{43}^{(k)}C_3^{(k)}e^{\beta_3^{(k)}r}+$$

$$+\mu_{44}^{(k)}C_4^{(k)}e^{\beta_4^{(k)}r}+\mu_{46}^{(k)}C_6^{(k)}e^{\beta_6^{(k)}r}+C_7^{(k)}\mu_{47}^{(k)}e^{\beta_7^{(k)}r}+C_8^{(k)}\mu_{48}^{(k)}e^{\beta_8^{(k)}r}+\bar{f}_4^{(k)}$$

$$f_7^{(k)}=C_7^{(k)}\mu_{77}^{(k)}e^{\beta_7^{(k)}r}+C_8^{(k)}\mu_{78}^{(k)}e^{\beta_8^{(k)}r}+\bar{f}_7^{(k)}$$

$$f_8^{(k)}=C_7^{(k)}\mu_{87}^{(k)}e^{\beta_7^{(k)}r}+C_8^{(k)}\mu_{88}^{(k)}e^{\beta_8^{(k)}r}+\bar{f}_8^{(k)} \qquad (12)$$

The integration constants $C_i^{(k)}$ are found from the contact conditions of the layers. For more complex cases of loading and boundary conditions, the following approximation of the sought functions is used:

$$U_x^{(k)}(x,r)=V_{11}^{(k)}(x)f_{11}^{(k)}(r)+V_{12}^{(k)}(x)f_{12}^{(k)}(r)$$

$$U_r^{(k)}(x,r)=V_{31}^{(k)}(x)f_{31}^{(k)}(r)+V_{32}^{(k)}(x)f_{32}^{(k)}(r)$$

$$\sigma_{xr}^{(k)}(x,r)=V_{41}^{(k)}(x)f_{41}^{(k)}(r)+V_{42}^{(k)}(x)f_{42}^{(k)}(r)$$

$$\sigma_{rr}^{(k)}(x,r)=V_{61}^{(k)}(x)f_{61}^{(k)}(r)+V_{62}^{(k)}(x)f_{62}^{(k)}(r)$$

$$T^{(k)}(x,r)=V_{71}^{(k)}(x)f_{71}^{(k)}(r)+V_{72}^{(k)}(x)f_{72}^{(k)}(r)$$

$$Q_3^{(k)}(x,r)=V_{81}^{(k)}(x)f_{81}^{(k)}(r)+V_{82}^{(k)}(x)f_{82}^{(k)}(r)$$

$$q_t^{(k)}(x,r)=V_{91}^{(k)}(x)q_1^{(k)}(r)+V_{92}^{(k)}(x)q_2^{(k)}(r) \qquad (13)$$

where $V_{i1}^{(k)}(x)=1-x/a$ and $V_{i2}^{(k)}(x)=x/a$. We obtain the equilibrium equations of a semi analytical finite element using the Reissner's variational principle:

$$[k_{00}]\{f_{1,3}^{(k)}\}=-[k_{01}]\{f_3^{(k)}\}+[k_{00}]/G_{13}^{(k)}\{f_4^{(k)}\},$$

$$[k_{00}]\{f_{3,3}^{(k)}\}=-B_{13}^{(k)}[k_{01}]\{f_1^{(k)}\}-\frac{1}{r^{(k)}}B_{23}^{(k)}[k_{00}]\{f_3^{(k)}\}+B_{33}^{(k)}[k_{00}]\{f_6^{(k)}\}+$$

$$+(B_{13}^{(k)}\alpha_1^{(k)}+B_{23}^{(k)}\alpha_2^{(k)}+\alpha_3^{(k)})[k_{00}]\{f_7^{(k)}\},$$

$$[k_{00}]\{f_{4,3}^{(k)}\} = B_{11}^{(k)}[k_{11}]\{f_1^{(k)}\} + \frac{1}{r^{(k)}} B_{12}^{(k)}[k_{10}]\{f_3^{(k)}\} - \frac{1}{r^{(k)}}[k_{00}]\{f_4^{(k)}\} + B_{13}^{(k)}[k_{00}]\{f_6^{(k)}\}$$

$$-(\alpha_1 B_{11}^{(k)} + \alpha_2 B_{12}^{(k)})[k_{01}]\{f_7^{(k)}(z)\}$$

$$[k_{00}]\{f_{6,3}^{(k)}\} = \frac{1}{r^{(k)}} B_{12}^{(k)}[k_{01}]\{f_1^{(k)}\} + \frac{1}{r^{(k)2}} B_{22}^{(k)}[k_{00}]\{f_3^{(k)}\} + [k_{10}]\{f_4^{(k)}\} + (\frac{1}{r^{(k)}} B_{23}^{(k)} - \frac{1}{r^{(k)}})[k_{00}]\{f_6^{(k)}\}$$

$$-\frac{1}{r}(\alpha_1 B_{21}^{(k)} + \alpha_2 B_{22}^{(k)})[k_{00}]\{f_7^{(k)}\}$$

$$[k_{00}]\{f_{7,3}^{(k)}\} = -[k_{00}]/\lambda_{rr}^{(k)}\{f_8^{(k)}\},$$

$$[k_{00}]\{f_{8,3}^{(k)}\} = \lambda_{xx}^{(k)}[k_{11}]\{f_7^{(k)}\} - \frac{1}{r^{(k)}}[k_{00}]\{f_8^{(k)}\} + [k_{00}]\{q^{(k)}\} \qquad (14)$$

where $[k_{00}] = \begin{bmatrix} a/3 & a/6 \\ a/6 & a/3 \end{bmatrix}$, $[k_{10}] = \begin{bmatrix} -1/2 & -1/2 \\ 1/2 & 1/2 \end{bmatrix}$ $[k_{11}] = \begin{bmatrix} 1/a & -1/a \\ -1/a & 1/a \end{bmatrix}$

$$[k_{01}] = [k_{10}]^T, \{f_i^{(k)}\} = \{f_{i1}^{(k)} \quad f_{i2}^{(k)}\}^T,$$

$$\{f_{i,3}^{(k)}\} = \{f_{i1,3}^{(k)} \quad f_{i2,3}^{(k)}\}^T, \quad \{q_i^{(k)}\} = \{q_{i1}^{(k)} \quad q_{i2}^{(k)}\}^T.$$

Next, we set up the governing system of differential equations for a layer subject to kinematic boundary conditions:

$$[K_{00}]\{f_{1j,3}^{(k)}\} = -[K_{01}]\{f_{3j}^{(k)}\} + [K_{00}]/G_{13}^{(k)}\{f_{4j}^{(k)}\},$$

$$[K_{00}]\{f_{3j,3}^{(k)}\} = -B_{13}^{(k)}[K_{01}]\{f_{1j}^{(k)}\} - \frac{1}{r^{(k)}} B_{23}^{(k)}[K_{00}]\{f_{3j}^{(k)}\} + B_{33}^{(k)}[K_{00}]\{f_{6j}^{(k)}\} +$$

$$+(B_{13}^{(k)}\alpha_1^{(k)} + B_{23}^{(k)}\alpha_2^{(k)} + \alpha_3^{(k)})[K_{00}]\{f_{7j}^{(k)}\},$$

$$[K_{00}]\{f_{4j,3}^{(k)}\} = B_{11}^{(k)}[K_{11}]\{f_{1j}^{(k)}\} + \frac{1}{r^{(k)}} B_{12}^{(k)}[K_{10}]\{f_{3j}^{(k)}\} - \frac{1}{r^{(k)}}[K_{00}]\{f_{4j}^{(k)}\} + B_{13}^{(k)}[K_{00}]\{f_{6j}^{(k)}\}$$

$$-(\alpha_1 B_{11}^{(k)} + \alpha_2 B_{12}^{(k)})[K_{01}]\{f_{7j}^{(k)}(z)\},$$

$$[K_{00}]\{f_{6j,3}^{(k)}\} = \frac{1}{r^{(k)}} B_{12}^{(k)}[K_{01}]\{f_{1j}^{(k)}\} + \frac{1}{r^{(k)2}} B_{22}^{(k)}[K_{00}]\{f_{3j}^{(k)}\} + [K_{10}]\{f_{4j}^{(k)}\} +$$

$$+(\frac{1}{r^{(k)}} B_{23}^{(k)} - \frac{1}{r^{(k)}})[K_{00}]\{f_{6j}^{(k)}\} - \frac{1}{r^{(k)}}(\alpha_1 B_{21}^{(k)} + \alpha_2 B_{22}^{(k)})[K_{00}]\{f_{7j}^{(k)}\},$$

$$[K_{00}]\{f_{7j,3}^{(k)}\} = -[K_{00}]/\lambda_{rr}^{(k)}\{f_{8j}^{(k)}\},$$

$$[K_{00}]\{f^{(k)}_{8j,3}\} = \lambda^{(k)}_{xx}[K_{11}]\{f^{(k)}_{7j}\} - \frac{1}{r^{(k)}}[K_{00}]\{f^{(k)}_{8j}\} + [K_{00}]\{q^{(k)}_j\} \qquad (15)$$

where $\{f^{(k)}_{ij}\} = \left\{f^{(k)}_{i1} \dots \dots f^{(k)}_{ij} \dots \dots f^{(k)}_{iJ}\right\}^T$, $\{f^{(k)}_{ij,3}\} = \left\{f^{(k)}_{i1,3} \dots \dots f^{(k)}_{ij,3} \dots \dots f^{(k)}_{iJ,3}\right\}^T$,

$\{q^{(k)}_j\} = \left\{q^{(k)}_1 \dots \dots q^{(k)}_j \dots \dots q^{(k)}_J\right\}^T$, j is the number of the point at which the unknown functions are determined. In this case, a partial solution of the steady-state heat conduction equation has the form:

$$\{\bar{f}^{(k)}_{7j}\} = (\lambda^{(k)}_{xx}[K_{11}])^{-1}[K_{00}]\{q^{(k)}_j\}, \{\bar{f}^{(k)}_{8j}\} = 0.$$

We obtain a partial solution of the system of differential Equations (15) from the solution of the following system of algebraic equations:

$$-[K_{01}]\{\bar{f}^{(k)}_{3j}\} + [K_{00}]/G^{(k)}_{13}\{\bar{f}^{(k)}_{4j}\} = 0,$$

$$-B^{(k)}_{13}[K_{01}]\{\bar{f}^{(k)}_{1j}\} - \frac{1}{r^{(k)}}B^{(k)}_{23}[K_{00}]\{\bar{f}^{(k)}_{3j}\} + B^{(k)}_{33}[K_{00}]\{\bar{f}^{(k)}_{6j}\} =$$

$$= -(B^{(k)}_{13}\alpha^{(k)}_1 + B^{(k)}_{23}\alpha^{(k)}_2 + \alpha^{(k)}_3)[K_{00}]\{\bar{f}^{(k)}_{7j}\},$$

$$B^{(k)}_{11}[K_{11}]\{\bar{f}^{(k)}_{1j}\} + \frac{1}{r^{(k)}}B^{(k)}_{12}[K_{10}]\{\bar{f}^{(k)}_{3j}\} - \frac{1}{r^{(k)}}[K_{00}]\{\bar{f}^{(k)}_{4j}\} + B^{(k)}_{13}[K_{00}]\{\bar{f}^{(k)}_{6j}\} =$$

$$= (\alpha_1 B^{(k)}_{11} + \alpha_2 B^{(k)}_{12})[K_{01}]\{\bar{f}^{(k)}_{7j}(z)\},$$

$$\frac{1}{r^{(k)}}B^{(k)}_{12}[K_{01}]\{\bar{f}^{(k)}_{1j}\} + \frac{1}{r^{(k)2}}B^{(k)}_{22}[K_{00}]\{\bar{f}^{(k)}_{3j}\} + [K_{10}]\{\bar{f}^{(k)}_{4j}\} +$$

$$+ \left(\frac{1}{r^{(k)}}B^{(k)}_{23} - \frac{1}{r^{(k)}}\right)[K_{00}]\{\bar{f}^{(k)}_{6j}\} = \frac{1}{r^{(k)}}(\alpha_1 B^{(k)}_{21} + \alpha_2 B^{(k)}_{22})[K_{00}]\{\bar{f}^{(k)}_{7j}\} \qquad (16)$$

Now, the solution of the system of differential Equations (15) can be written as follows:

$$\begin{bmatrix} \{f_{1j}^{(k)}\} \\ \{f_{3j}^{(k)}\} \\ \{f_{4j}^{(k)}\} \\ \{f_{6j}^{(k)}\} \\ \{f_{7j}^{(k)}\} \\ \{f_{8j}^{(k)}\} \end{bmatrix} = \begin{bmatrix} \mu_{1j}^{(k)}(1) & ,..., & \mu_{1j}^{(k)}(j) & ,..., & \mu_{11}^{(k)}(J) \\ \mu_{3j}^{(k)}(1) & ,..., & \mu_{3j}^{(k)}(j) & ,..., & \mu_{3j}^{(k)}(J) \\ \mu_{4j}^{(k)}(1) & ,..., & \mu_{4j}^{(k)}(j) & ,..., & \mu_{4j}^{(k)}(J) \\ \mu_{6j}^{(k)}(1) & ,..., & \mu_{6j}^{(k)}(j) & ,..., & \mu_{6j}^{(k)}(J) \\ \mu_{7j}^{(k)}(1) & ,..., & \mu_{7j}^{(k)}(j) & ,..., & \mu_{7j}^{(k)}(J) \\ \mu_{8j}^{(k)}(1) & ,..., & \mu_{8j}^{(k)}(j) & ,..., & \mu_{8j}^{(k)}(J) \end{bmatrix} \left[C^{(k)}\right] + \begin{bmatrix} \{\bar{f}_{1j}^{(k)}\} \\ \{\bar{f}_{3j}^{(k)}\} \\ \{\bar{f}_{4j}^{(k)}\} \\ \{\bar{f}_{6j}^{(k)}\} \\ \{\bar{f}_{7j}^{(k)}\} \\ \{\bar{f}_{8j}^{(k)}\} \end{bmatrix} \quad (17)$$

where $\left[C^{(k)}\right]^T = \left[C_1^{(k)} e^{r\beta_1^{(k)}}, ..., C_j^{(k)} e^{r\beta_j^{(k)}}, ..., C_J^{(k)} e^{r\beta_J^{(k)}}\right]$ $\beta_j^{(k)}$ - are the roots of the characteristic Equation of the governing system of differential equations; $\mu_{1j}^{(k)}(j)$ $\mu_{3j}^{(k)}(j)$ $\mu_{4j}^{(k)}(j)$ $\mu_{6j}^{(k)}(j)$ $\mu_{7j}^{(k)}(j)$ $\mu_{8j}^{(k)}(j)$ are its eigenvectors; $C_j^{(k)}$ - are the constants of integrations determined from the interface conditions between the layers and the boundary conditions on the outside surfaces at each node of the finite-element partitions of the structure; J is the total of unknown functions in a layer.

To test the proposed approach, a model with a polynomial approximation of the sought displacements along the thickness of the shell has been developed. We use the following approximation of displacements Marchuk et al. [7]:

$$U_x^{(k)}(x,r) = U_{xs}^{(k)}(x)\xi_s^{(k)}(r) + \frac{\partial}{\partial x} W_p^{(k)}(x)\phi_p^{(k)}(r),$$

$$U_r^{(k)}(x,r) = W_p^{(k)}(x)\psi_p^{(k)}(r) \quad (18)$$

Here $U_{x1}^{(k)}(x)$ and $U_{x2}^{(k)}(x)$ are the tangential displacements on the face surfaces of the construction along the x axis; $W_1^{(k)}(x)$ and $W_2^{(k)}(x)$ are the normal displacements on the face surfaces of the construction; $W_3^{(k)}(x)$ is the shifting function; $\xi_1^{(k)}(r)$ $\xi_2^{(k)}(r)$ $\psi_1^{(k)}(r)$ $\psi_2^{(k)}(r)$ and are the given polynomials of the first degree; $\varphi_1^{(k)}(r)$ $\varphi_2^{(k)}(r)$ $\psi_3^{(k)}(r)$ and some of the second degree, and $\varphi_3^{(k)}(r)$ some of the third degree. We do not present the derivation of the differential equations for shell equilibrium under temperature influence to reduce the article's volume. This is similar to that given in Marchuk et al. [7, 8].

Calculation of Results

As the first test example, we will consider the thermo-elastic equilibrium of a three-layer isotropic cylindrical shell hinge-supported along its contour. The physicomechanical characteristics of the load-carrying layers and filler are as follows: $E^{(1)} = E^{(3)} = E_0$ $E_0 = 1$ MPa, $E^{(1)} / E^{(2)} = 100$, $\nu^{(1)} = \nu^{(2)} = \nu^{(3)} = 0.3$, $G^{(1)} = E^{(1)} / [2*(1+\nu^{(1)})]$, $G^{(2)} = E^{(2)} / [2*(1+\nu^{(2)})]$, $G^{(3)} = G^{(1)}$ $h = h_0 = 1$ $h^{(2)} = h/2$ $h^{(1)} = h^{(3)} = h/4$ $h^{(2)} = h/2$ m, $l/h = 10$, $R/h = 3$, $R = r^{(2)}$, $\lambda^{(1)} = \lambda^{(3)} = 1$ W/(m deg), $\lambda^{(1)} / \lambda^{(2)} = 100$ $\alpha_0 = 1$/deg, $\alpha^{(1)} = \alpha^{(3)} = 23.8*10^{-6}*$ α_0, $\alpha^{(1)} / \alpha^{(2)} = 100$ The contact of layers is rigid. The thermo mechanical contact of the layers is ideal. The shell is loaded on the inner and outer surface at sinusoidal temperature $T^{(1)} = 50*T_0$ $T^{(3)} = 0*T_0$ $T_0 = 1$, deg. Calculations were carried out by the analytical model (A) and the model with a polynomial approximation of sought-for functions across the layer thickness (P) without splitting the layers into sub layers. In addition, when calculated by the analytical model, each layer was divided into four sub-layers (A_4). Also, the calculation was carried out using a model without taking into account transverse compression (S) [performed by appropriately clarifying the physical and mechanical characteristics when calculating using model (A)].

Using the analytical model (A_4), the temperature values at the boundary of the layers are 50.0, 49.636, 0.2175, and 0. Table 1 lists the dimensionless on-layer border displacements $\bar{U}_1 = U_1(0,r)/(h_0\alpha_0 T_0*10^{-3})$ $\bar{U}_3 = U_3(l/2,r)/(h_0\alpha_0 T_0*10^{-3})$. Stress distributions ($\bar{\sigma}_{11} = \sigma_{11}(l/2,r)/(\alpha_0 T_0 E_0)*10^{-6}$ and $\bar{\sigma}_{22} = \sigma_{22}(l/2,r)/(\alpha_0 T_0 E_0)*10^{-6}$) on layer borders are presented in Table 2.

Table 1. Dimensionless displacements at the interfaces of isotropic cylindrical shell layers under the influence of temperature load

No. of the layer	$\bar{U}_1$				$\bar{U}_3$			
	P	A	S	A_4	P	A	S	A_4
1	-3.0542 -3.2254	-3.0526 -3.2235	-3.1499 -3.2714	-3.0509 -3.2221	2.1108 2.4578	2.1093 2.4564	1.6667 1.6667	2.1120 2.4592
2	-.32254 -.48734	-3.2235 -4.8886	-3.2714 -.27886	-3.2221 -.48906	2.4578 1.0022	2.4564 1.0048	1.6667 1.6667	2.4592 1.0058
3	-.48734 -.55685	-4.8886 -5.5865	-.27886 -4.0232	-.48906 -.55893	1.0022 .95214	1.0048 9.5466	1.6667 1.6667	1.0058 .95562

Table 2. Dimensionless stresses at the interfaces of isotropic cylindrical shell layers under the influence of temperature load

No. of the layer	$\bar{\sigma}_{11}$				$\bar{\sigma}_{22}$			
	P	A	S	A_4	P	A	S	A_4
1	-367.24 -292.66	-368.00 -293.47	-392.77 -355.82	-368.22 -293.65	-455.86 -384.61	-456.66 -385.35	-641.15 -680.16	-455.65 -384.42
2	-1.8805 -9.6328	-1.9071 -9.6494	13.123 2.6526	-1.9538 -9.6706	-2.8000 -8.4384	-2.8259 -8.4525	9.8798 5.9236	-2.8615 -8.4719
3	250.18 281.93	250.95 282.78	257.95 295.88	251.09 282.97	369.61 356.62	370.64 357.59	585.05 564.97	370.95 357.93

The calculation results using the analytical model (A) and the model with a polynomial approximation of the desired functions over the layer thickness (P) practically coincide. Splitting layers into sub layers does not significantly refine the calculation results (A_4). The given comparisons have shown the inappropriateness of applying shear models (S) in such problems.

Consider the same shell. The first layer has an internal heat source $q_t^{(1)} = 10 * q_0$, q_0 =1W/m³ ($T^{(1)} = 0 * T_0, T^{(3)} = 0 * T_0$). The temperature values at the boundaries of the layers using the analytical model (A_4) are 0, 3.1998,0.0140, and 0.

Table 3 lists the dimensionless displacements $\bar{U}_1 = U_1(0,r)\lambda_0 /(h_0^3 \alpha_0 q_0 * 10^{-5})$ and $\bar{U}_3 = U_3(l/2,r)\lambda_0 /(h_0^3 \alpha_0 q_0 * 10^{-5})$ stress distributions ($\bar{\sigma}_{11} = \sigma_{11}(l/2,r)\lambda_0 /(h_0^2 \alpha_0 q_0 E_0) * 10^{-7}$ and $\bar{\sigma}_{22} = \sigma_{22}(l/2,r)\lambda_0 /(h_0^2 \alpha_0 q_0 E_0) * 10^{-7}$) on layer borders. Table 4 presents these data.

Table 3. Dimensionless displacements at the interfaces of isotropic cylindrical shell layers in the presence of an internal heat source

No. of the layer	$\bar{U}_1$				$\bar{U}_3$			
	P	A	S	A_4	P	A	S	A_4
1	-13.248 -14.043	-13.202 -13.993	-13.676 -14.259	-13.254 -14.034	9.5217 11.031	9.4851 10.991	7.3637 7.3637	9.3417 10.852
2	-14.043 -2.1078	-13.993 -2.1082	-14.259 -1.2107	-14.034 -2.1320	11.031 4.5394	10.991 4.5363	7.3637 7.3637	10.852 4.4640
3	-2.1078 -2.4239	-2.1082 -2.4245	-1.2107 -1.7564	-2.1320 -2.4428	4.5394 4.3184	4.5363 4.3155	7.3637 7.3637	4.4640 4.2451

Table 4. Dimensionless stresses at the interfaces of isotropic cylindrical shell layers in the presence of an internal heat source

No. of the layer	$\bar{\sigma}_{11}$				$\bar{\sigma}_{22}$			
	P	A	S	A_4	P	A	S	A_4
1	582.93 -482.11	580.86 -480.14	569.23 -507.39	580.76 -480.33	555.75 -512.90	553.66 -510.87	465.32 -645.99	547.89 -515.90
2	-.99794 -4.3397	-1.0055 -4.3331	5.6966 1.1644	-.94442 -4.2687	-1.3059 -3.7746	-1.3128 -3.7689	4.3105 2.6148	-1.3002 -3.7274
3	108.49 124.36	108.51 124.35	111.72 130.00	108.65 124.32	165.00 160.69	164.93 160.61	256.76 249.39	162.79 158.58

As in the previous case, the calculation results using the analytical model (A) and the model with a polynomial approximation of the desired functions over the layer thickness (P) practically coincide. The splitting of layers into sub-layers does not significantly refine the calculation results (A). The comparisons show the inappropriate application of shear models (S) in such problems. As a second test example, let us consider the thermal stress state of a four-layer orthotropic cylindrical shell, hinge-supported along its contour. The physical and mechanical characteristics of the layers are as follows: 18.2*1000* E_0, $E_2^{(1)} = 41.5*1000*E_0$, $E_3^{(1)} = E_1^{(1)}$ $G_{12}^{(1)} = G_{13}^{(1)} = G_{23}^{(1)}$ =6.83*1000* E_0, $E_0 = 1$ MPa, $\nu_{12}^{(1)} = \nu_{13}^{(1)} = \nu_{32}^{(1)} = 0.257$ ($\nu_{21}/E_2 = \nu_{12}/E_1$), $\lambda_1^{(1)} = \lambda_3^{(1)}$=0.8 * λ_0, $\lambda_2^{(1)} = 1.2*\lambda_0$, $\lambda_0 = 1$ W/(m deg) $\alpha_1^{(1)} = \alpha_3^{(1)}$=16.3*$10^{-6}$* α_0, $\alpha_2^{(1)} = 8.0*10^{-6}*\alpha_0$, α_0=1/deg $h^{(1)} = h/4$, The second and fourth layers are identical to the first, but turned around 90^0; $E_1^{(3)}$=30.1*1000* E_0, $E_2^{(3)}$ = 26.5*1000* E_0, $E_3^{(3)} = E_2^{(3)}$ $G_{12}^{(3)} = G_{13}^{(3)} = G_{23}^{(3)}$=4.28*1000* E_0, $\nu_{12}^{(3)} = \nu_{13}^{(3)} = 0.12267$,

$\nu_{32}^{(3)} = 0.108$, $\lambda_1^{(3)} = 1 * \lambda_0$, $\lambda_2^{(3)} = \lambda_3^{(3)} = 0.4 * \lambda_0$, $\alpha_1^{(3)} = 12.7 * 10^{-6} * \alpha_0$, $\alpha_1^{(3)} = \alpha_3^{(3)} = 14.37 * 10^{-6} * \alpha_0$, $h^{(3)} = h/4$. The thermo-mechanical contact of the layers is ideal. The shell is loaded on the inner and outer surfaces with sinusoidal temperature $T^{(1)} = 50 * T_0$ $T^{(4)} = 0 * T_0$. The temperature value at the boundaries of the layers using the analytical model (A_4): 50.0, 38.159, 27.629, 8.6895, and 0.

Table 5 lists the dimensionless displacements $\bar{U}_1 = U_1(0,r)/(h_0 \alpha_0 T_0 * 10^{-4})$ and $\bar{U}_3 = U_3(l/2,r)/(h_0 \alpha_0 T_0 * 10^{-4})$ stress distributions ($\bar{\sigma}_{11} = \sigma_{11}(l/2,r)/(\alpha_0 T_0 E_0)$ and $\bar{\sigma}_{22} = \sigma_{22}(l/2,r)/(\alpha_0 T_0 E_0)$) on layer borders. Table 6 presents these data.

Table 5. Dimensionless displacements at the interfaces of orthotropic cylindrical shell layers under the influence of temperature load

No. of the layer	$\bar{U}_1$				$\bar{U}_3$			
	P	A	S	A_4	P	A	S	A_4
1	-6.3851	-6.3909	-6.2024	-6.3628	6.8095	6.8099	9.4209	6.8751
	-6.8488	-6.8524	-6.7769	-6.8298	9.2682	9.2677	9.4209	9.3304
2	-6.8488	-6.8524	-6.7769	-6.8298	9.2682	9.2677	9.4209	9.3304
	-7.2480	-7.2520	-7.1475	-7.2348	10.834	10.833	9.4209	10.895
3	-7.2480	-7.2520	-7.1475	-7.2348	10.834	10.833	9.4209	10.895
	-7.4291	-7.4358	-7.2360	-7.4234	11.368	11.367	9.4209	11.429
4	-7.4291	-7.4358	-7.2360	-7.4234	11.368	11.367	9.4209	11.429
	-8.0887	-8.0984	-7.7780	-8.0911	10.747	10.748	9.4209	10.808

Table 6. Dimensionless stresses at the interfaces of orthotropic cylindrical shell layers under the influence of temperature load

No. of the layer	$\bar{\sigma}_{11}$				$\bar{\sigma}_{22}$			
	P	A	S	A_4	P	A	S	A_4
1	-14.833	-14.829	-13.644	-14.815	-14.134	-14.131	-9.1022	-14.014
	-8.7400	-8.7378	-8.3479	-8.7253	-4.3686	-4.3678	-3.4553	-4.2667
2	-8.7926	-8.7869	-8.1628	-8.7944	-7.6590	-7.6580	-7.1977	-7.6190
	-1.7925	-1.7866	-1.6427	-1.7892	-2.3600	-2.3593	-2.9139	-2.3233
3	-4.0183	-4.0153	-4.1282	-4.0256	-1.4965	-1.4977	-2.6460	-1.4447
	4.4024	4.4093	4.1121	4.4033	6.3637	6.3642	4.8171	6.4136
4	10.011	10.023	9.5404	10.026	6.1869	6.1901	5.1469	6.2250
	16.271	16.286	15.319	16.297	9.7700	9.7743	8.8358	9.8086

As in the case of an isotropic shell, for an orthotropic shell, the results of calculations by models (A), (P) and (A_4) practically coincide. The calculation using the shear model leads to significant errors. The convergence of the proposed semi analytical finite element method is demonstrated in Table 7. We consider a four-layered orthotropic cylindrical shell as in the previous problem. We divide the half shell into 2 (MA_2), 4 (MA_4), 8 (MA_8), 16 (MA_16), and 32 (MA_32) finite elements along the x-axis; the layers were not split into sub layers. The shell is loaded on the inner and outer surface with sinusoidal heat flux $Q^{(1)}=10*Q_0$ $Q^{(4)}=0*Q_0$ Q_0=1 W/m^2. We consider displacements ($\bar{U}_1=U_1(0,r)*\lambda_0/(h_0^2\alpha_0Q_0*10^{-3})$ and $\bar{U}_3=U_3(l/2,r)*\lambda_0/(h_0^2\alpha_0Q_0*10^{-3})$ and stresses ($\bar{\sigma}_{11}=\sigma_{11}(l/2,r)*\lambda_0/(h_0\alpha_0Q_0E_0)$ $\bar{\sigma}_{22}=\sigma_{22}(l/2,r)*\lambda_0/(h_0\alpha_0Q_0E_0)$ on the inner and outer surfaces of the shell.

Table 7. Convergence of the proposed semi analytical finite element method

Displacements, stresses	MA_2	MA_4	MA_8	MA_16	MA_32	A
$\bar{U}_1$	-1.8990 -2.7284	-2.0885 -2.9423	-2.1378 -2.9976	-2.1502 -3.0115	-2.1533 -3.0150	-2.1543 -3.0162
$\bar{U}_3$	2.3058 3.5377	2.3648 3.6287	2.3802 3.6525	2.3841 3.6585	2.3851 3.6600	2.3854 3.6605
$\bar{\sigma}_{11}$	-13.282 6.3600	-12.126 11.226	-11.794 12.561	-11.708 12.903	-11.686 12.989	-11.679 13.017
$\bar{\sigma}_{22}$	3.4844 -1.7980	4.1529 -.94973	4.3488 -.70876	4.3997 -.64662	4.4125 -.63096	4.4168 -.62573
T	80.6439 73.4905	83.5963 76.4391	84.3574 77.1992	84.5490 77.3906	84.5970 77.4386	84.6131 77.4545

Completing half the shell into 8 (MA_8) finite elements along the x-axis achieves satisfactory calculation accuracy. Consider the same shell but with the first layer delaminated along a part of the shell length. Half of the shell is divided into 32 elements along the x-axis. At $<0<x<17*l/64$, there is no delamination at $17*l/64<x<l/2$. The development of delamination is not considered. The thermo mechanical contact of the layers is ideal. It is assumed that the distance between the delaminated layers is negligible and does not

affect the thermo mechanical contact conditions. The temperature and heat flow at the layer boundaries are inseparable. The shell is loaded on the inner and outer surfaces at sinusoidal temperature $T^{(1)} = -50*T_0$ $T^{(4)} = 0*T_0$. Table 8 and Table 9 list the dimensionless displacements $\bar{U}_1 = U_1/(h\alpha_0 T_0 *10^{-4})$ and $\bar{U}_3 = U_3/(h\alpha_0 T_0 *10^{-4})$, as well as stresses $\bar{\sigma}_{11} = \sigma_{11}/(\alpha_0 T_0 E_0)$ $\bar{\sigma}_{22} = \sigma_{22}/(\alpha_0 T_0 E_0)$ at the interfaces of the layers of the shell.

Table 8. Dimensionless displacements and stresses at layer boundaries of a delaminated shell ($x = l/2$)

No. of the layer	$\bar{U}_1$		$\bar{U}_3$		$\bar{\sigma}_{11}$		$\bar{\sigma}_{22}$	
	MA_32	MS_32	MA_32	MS_32	MA_32	MS_32	MA_32	MS_32
1	0	0	-11.251	-12.342	11.617	10.999	4.8761	2.7030
	0	0	-13.341	-12.342	5.6035	5.9256	-4.0695	-2.3728
2	0	0	-5.6744	-7.2611	9.0512	8.9472	9.9055	8.8286
	0	0	-7.4205	-7.2611	2.5867	2.4645	4.5139	4.4353
3	0	0	-7.4205	-7.2611	4.1886	4.3141	4.4800	4.3141
	0	0	-8.0169	-7.2611	-3.9439	-3.8259	-3.6023	-3.8259
4	0	0	-8.0169	-7.2611	-8.6457	-8.5982	-4.0072	-3.6953
	0	0	-7.5481	-7.2611	-14.176	-14.044	-7.5682	-7.3851

Table 9. Dimensionless displacements and stresses at layer boundaries of a delaminated shell ($x = 14l/64$)

No. of the layer	$\bar{U}_1$		$\bar{U}_3$		$\bar{\sigma}_{11}$		$\bar{\sigma}_{22}$	
	MA_3 2	MS_32	MA_32	MS_32	MA_32	MS_32	MA_32	MS_32
1	.3709	5.2297	-3.3356	-5.3758	8.3652	8.5889	9.9882	6.7326
	5.6550	5.5565	-4.9963	-5.3758	5.9096	5.7305	24.495	21.239
2	5.6550	5.5565	-4.9963	-5.3758	6.0715	5.8182	5.3688	5.1282
	5.7798	5.6855	-6.0648	-5.3758	1.1519	1.0035	13.275	13.034
3	5.7798	5.6855	-6.0648	-5.3758	2.4586	2.3649	1.6481	2.0180
	5.7539	5.6414	-6.4107	-5.3758	-2.9178	-2.5687	4.0059	4.2319
4	5.7539	5.6414	-6.4107	-5.3758	-6.3766	-6.1107	-3.4850	-2.9437
	6.0557	5.9090	-6.0180	-5.3758	-11.121	-10.064	-3.4850	-2.9437

The comparisons show the inappropriate application of shear models (MS_32) in such problems. Consider the same shell. Half of the shell is divided into 32 elements along the *x*-axis. At $0<x<17*l/64$ the outer surface of the shell, it is free (fourth layer), and $17*l/64<x<l/2$ radial displacements are forbidden on the outer surface. The shell is loaded on the inner and outer surface with sinusoidal temperature $T^{(1)}=50*T_0$ $T^{(4)}=0*T_0$; Table 10 and Table 11 list the dimensionless displacements $\bar{U}_1=U_1/(h\alpha_0T_0*10^{-4})$ and $\bar{U}_3=U_3/(h\alpha_0T_0*10^{-4})$, as well as stresses $\bar{\sigma}_{11}=\sigma_{11}/(\alpha_0T_0E_0)$, $\bar{\sigma}_{22}=\sigma_{22}/(\alpha_0T_0E_0)$ at the interfaces of the layers of the shell.

Table 10. Dimensionless displacements and stresses at the boundaries of shell layers with complex boundary conditions on the outer surface ($x=l/2$)

No. of the layer	$\bar{U}_1$		$\bar{U}_3$		$\bar{\sigma}_{11}$		$\bar{\sigma}_{22}$	
	MA_32	MS_32	MA_32	MS_32	MA_32	MS_32	MA_32	MS_32
1	0	0	-4.2498	0	-16.207	-15.291	-33.297	-25.706
	0	0	-1.1984	0	-10.737	-10.093	-22.320	-18.695
2	0	0	-1.1984	0	-6.8449	-6.6354	-14.518	-13.040
	0	0	.30436	0	-1.2858	-1.0510	-9.1566	-8.4771
3	0	0	.30436	0	-1.2617	-2.3315	-10.727	-10.774
	0	0	.68485	0	6.1829	5.1478	-2.4354	-2.7526
4	0	0	.68485	0	8.7874	9.0889	-.77865	-.24483
	0	0	0	0	13.725	13.951	2.7789	3.5854

The comparisons show the inappropriate application of shear models (MS_32) in such problems. Consider the stress-strain state of an orthotropic four-layer shell under the influence of a local heat flux. Half of the shell is divided into 100 elements along the x-axis. $90*l/200<x<l/2$ The shell on the inner surface is loaded with uniformly distributed heat flux $Q^{(1)}=10*Q_0$; there is no heat flux on the shell's outer surface $Q^{(4)}=0*Q_0$. The temperature values at the boundaries of the layers are 22.7124, 20.1727, 18.7319, 17.4006, and 17.1935 ($x=l/2$). Table 12 lists the dimensionless

$\bar{U}_1 = U_1(0,r) * \lambda_0 / (h_0^2 \alpha_0 Q_0 * 10^{-4})$ and $\bar{U}_3 = U_3(l/2,r) * \lambda_0 / (h_0^2 \alpha_0 Q_0 * 10^{-4})$

well stresses $\bar{\sigma}_{11} = \sigma_{11}(l/2,r) * \lambda_0 / (h_0 \alpha_0 Q_0 E_0 * 10^{-3})$ and $\bar{\sigma}_{22} = \sigma_{22}(l/2,r) * \lambda_0 / (h_0 \alpha_0 Q_0 E_0 * 10^{-3})$ at the interfaces of the layers of the shell.

Table 11. Dimensionless displacements and stresses at the boundaries of shell layers with complex boundary conditions on the outer surface

($x = 14l/64$)

No. of the layer	$\bar{U}_1$.		$\bar{U}_3$		$\bar{\sigma}_{11}$		$\bar{\sigma}_{22}$	
	MA_32	MS_32	MA_32	MS_32	MA_32	MS_32	MA_32	MS_32
1	-10.286 -9.5207	-8.9451 -8.3215	-.58923 1.2949	1.5797 1.5797	-11.469 -6.9495	-10.732 -6.4746	-18.322 -32.829	-14.290 -28.796
2	-9.5207 -8.8886	-8.3215 -7.7243	1.2949 2.3394	1.5797 1.5797	-5.9820 -.64422	-5.2959 -.63600	-8.1318 -16.038	-7.5063 -15.412
3	-8.8886 -8.3698	-7.7243 -7.1658	2.3394 2.6908	1.5797 1.5797	-1.4461 3.3611	-1.6992 3.1981	-4.8646 -7.2224	-5.4631 -7.8209
4	-8.3698 -7.8194	-7.1658 -6.6177	2.6908 2.3536	1.5797 1.5797	6.2198 12.872	6.0116 11.090	1.3482 1.3482	.79244 .79244

Table 12. The stress-strain state of an orthotropic four-layer shell under the influence of a local heat flux

No. of the layer	$\bar{U}_1$		$\bar{U}_3$		$\bar{\sigma}_{11}$		$\bar{\sigma}_{22}$	
	MA_100	MS_100	MA_1 00	MS_10 0	MA_10 0	MS_100	MA_100	MS_100
1	-4.2114 -4.5356	-4.2589 -4.6288	4.7523 5.7442	6.2327 6.2327	-3611.2 -2379.7	-3186.6 -2130.4	-1833.9 533.24	872.42 1401.4
2	-4.5356 -4.9003	-4.6288 -5.0031	5.7442 6.5807	6.2327 6.2327	-131.46 996.00	157.44 1077.1	-2217.3 - 1320.0	-1826.4 -1505.8
3	4.9003 -5.2939	-5.0031 -5.3714	6.5807 7.2462	6.2327 6.2327	-1503.9 -671.11	-1407.6 -679.56	-1484.5 - 792.45	-1779.6 -1617.6
4	-5.2939 -5.7330	-5.3714 -5.7398	7.2462 7.8580	6.2327 6.2327	2225.0 3632.0	1997.5 2808.9	-543.91 - 87.299	-1164.7 -1144.0

The comparisons show the inappropriate application of shear models (MS_100) in such problems.

Discussion

For the first time in this chapter, a three-dimensional analytical solution is presented for a simply supported, layered orthotropic cylindrical shell under the assumption of a constant layer radius. This approximation was found to provide accurate results with minimal error. The solution addresses scenarios involving thermal loading, heat flow, and internal heat sources, serving as a valuable benchmark for future research on alternative calculation methods.

This chapter presents a compelling case against the use of simplified shear calculation models in the analysis of layered composite cylindrical shells under thermal loads, demonstrating their inadequacy in accurately modelling the complex stress-strain state of these structures. Including an analytical approach to internal heat sources in the presented solution also holds potential for future dynamic analysis of dissipative heating in layered shells.

A new version of the semi analytical finite element method has also been developed to solve problems of thermal loading in layered cylindrical shells. In the classic version of the semi analytical finite element method, approximation by trigonometric series is used for one of the coordinates, and polynomials for the others. The semi analytical finite element method presented in this chapter approximates the required functions in the shell plane using polynomials. It obtains its thickness distribution based on the analytical solution of the corresponding system of differential equations.

This approach enables three-dimensional analysis of a layered composite cylindrical shell subject to local temperature loads, partial delamination along the shell plane, and complex boundary conditions on the shell's surface, overcoming significant challenges in modelling such scenarios.

The semi analytical finite element method presented in this chapter holds significant potential for future research in the thermo-elastic analysis of layered composite cylindrical shells. The technique could be extended to analyze delamination under unsteady thermal loading and friction-induced heating in dynamic problems—issues that can only be adequately addressed using three-dimensional solutions. These advancements would provide critical insights into the complex thermal behaviour of layered structures in various practical applications.

Conclusion

This chapter introduces a three-dimensional analytical solution for a supported, layered orthotropic cylindrical shell, using a constant layer radius approximation that yields accurate results. The method effectively addresses thermal loading, heat flow, and internal heat sources, challenging the use of simplified shear models for these structures. Additionally, a novel semi analytical finite element method is developed for detailed three-dimensional thermo-elastic analysis, providing a promising foundation for future studies on delamination, dynamic thermal loads, and dissipative heating in layered shells.

Disclaimer

None.

References

[1] Ootao Y., Tanigawa Y. & Miyatake K. (2010). Transient thermal stresses of the cross-ply laminated cylindrical shell using a higher-order shear deformation theory, *Journal of Thermal Stresses,* 33(1), 55–74.

[2] Brischetto S. & Carrera E. (2011). Heat conduction and thermal analysis in multilayered plates and shells, *Mechanics Research Communications,* 38(6), 449–455.

[3] Tokovyy Y., Chyzh A. & Ma C. C. (2019). An analytical solution to the asymmetric thermoelasticity problem for a cylinder with arbitrarily varying thermomechanical properties, *Acta Mechanica*, 230, 1469–1485.

[4] Musii R., Zhydyk U., Svidrak I., Shynder V. & Morska N. (2023). Determination and analysis of the thermo-elastic state of layered orthotropic cylindrical shells, *Mathematical Modeling and Computing,* 10(3), 918–926.

[5] Grigorenko, J. M., Vasilenko, A. T., & Pankratova N. D. (1991). Zadachi of the theory of elasticity of non-uniform ph., *Naukova Dumka,* Kyiv.

[6] Grigorenko J. M., Vlaikov G. G., & Grigorenko A. J. (2006). Numerical-Analytical Solution of Problems of Shell Mechanics Based on Different Models [in Russian] ph. Akademperiodika, Kyiv.

[7] Marchuk, A. V., &Putvinskayte, Yu. K. (2019). Analytical solution of the problem on the thermally stressed state of composite plates with rigid and sliding contacts betweenlayersbased on the 3d elasticity theory, *Mechanics of Composite Materials,* 55(2), 155–170.

[8] Marchuk, A. V. (2024). Analytical solution to the three-dimensional problem on the thermal stress of composite Plates and Shallow Shells in the Presence of an internal heat Source, New Research on Thermal Stresses, Nova Science Publisher Inc. New York, 2, 45-59.

[9] Marchuk A. V., Gnedash S. V., & Shandyba D. O. (2017). Free and forced vibrations of thick-walled laminated anisotropic cylindrical shells account for energy dissipation at frequencies close to resonance ones, *Composite Structures: Mechanics, Computation, and Applications: An International Journal*, 8(3), 239-265.

[10] Marchuk A. V., & Gnedash S. V. (2016). Analysis of the effect of local loads on thick-walled cylindrical shells with different boundary conditions, *Int. Applied Mechanics,* 52(4), 368-376.

[11] Marchuk A. V., Gnedash S. V., Apunevich A. A., & Vovk A. V. (2017). A study of the effect of friction between delaminated layers on the stress-strain state of thick laminated anisotropic cylindrical shells by the semi analytical finite element method, *Strength of Materials,* 49(5), 627-634.

Chapter 4

Analysis of Circular Thickness and Bi-Linear Thermal Gradient Impact on Period of Parallelogram-Shaped Plates

Neeraj Lather[1], PhD
Sapna[1], MSc
Naveen Mani[2], PhD
Reeta Bhardwaj[1], PhD
Sudeshna Ghosh[1], PhD
and Amit Sharma[1,*], PhD
[1]Department of Mathematics, Amity School of Applied Sciences, Amity University Haryana, Haryana, India
[2]Department of Mathematics, Chandigarh University, Mohali, Punjab, India

Abstract

The current research considers a plate with a parallelogram shape, featuring a circular profile for thickness and density in two and one dimensions, respectively. Additionally, a linear thermal gradient profile is considered. The plate has different edge conditions: SCCC and CSCC. The Rayleigh-Ritz method was used to solve the frequency equations. The main conclusions drawn from the study are as follows: the frequency obtained for the system was lower than that of the previously studied system, and the system converged to a stable state very rapidly.

Keywords: vibration, plate, thermal gradient, circular thickness, density

---

* Corresponding Author's Email: dba.amitsharma@gmail.com

In: Thermal Modeling Reimagined
Editors: Pankaj Thakur and Jatinder Kaur
ISBN: 979-8-89530-463-1

Introduction

In the current scenario, with advancements in economic and scientific technology, plates with non-homogeneous configurations are widely used in structural design. Compared to homogeneous plates, non-homogeneous plates offer greater durability. Due to these advantages, their use has been steadily increasing. In particular, parallelogram plates are extensively used in the design of automobiles, marine structures, aircraft, and more. This practical applicability of parallelogram plates prompted the current research.

Understanding their vibration characteristics is essential for designing such structures to ensure efficiency, accuracy, and safety. This necessity leads to a thorough study of the free vibration of parallelogram plates. The first few vibration modes play a significant role in understanding the vibration characteristics of the plate. Numerous research articles have been published on this topic. Pagani et al. [1] developed a novel numerical approach to study the vibrational behavior of variable angle tow composite structures in their quasi-static nonlinear equilibrium states. A hybrid quasi-3D theory by Vinh [2] was created to study deflections, stresses, and vibrations in bi-functionally graded sandwich plates on Pasternak's elastic foundations. Farag & Ashour [3] examined vibrational effects on thin orthotropic skew plates using a semi-analytical method, focusing on the influence of skew angle, aspect ratio, and edge conditions. Sharma & Sharma [4] investigated vibrational effects on non-homogeneous parallelogram plates with clamped boundaries, considering 2D linearly varying temperature and bilinear thickness.

Sharma & Dhiman [5] analyzed vibrations of orthotropic parallelogram plates with linear temperature distribution and bi-parabolic thickness variation in both directions, using the variable separable technique to solve the system's equations. Gupta et al. [6] calculated the period and deflection functions for the first two vibration modes of viscoelastic orthotropic parallelogram plates with clamped boundaries, varying taper constants, aspect ratios, and skew angles. The Rayleigh-Ritz technique was applied by Gupta et al. [7] to analyze the vibration of clamped orthotropic parallelogram plates with linear variations in thickness. Sharma & Sharma [8] computed the natural vibration of parallelogram plates with variations in thickness and thermal effects using the variable separable method.

Kumar et al. [9] analyzed the effects of linear density fluctuations and circular variations in Poisson's ratio on the vibration periods of tapered rectangular plates in a temperature field. Conway & Farnham [10] modified the point-matching approach to examine the fundamental free flexural vibrations of supporte disosceles triangular, rhombic, and parallelogram plates. Lather & Sharma [11] used the Rayleigh-Ritz technique to solve the differential equations for the first four vibration modes of rectangular orthotropic plates with circular variations in density and thickness. Lather & Sharma [12] also examined the time period of an isotropic clamped rectangular plate with varying thickness and temperature. Kumar et al. [13] studied the natural vibrations of an isotropic viscoelastic square plate with circular thickness changes and Poisson's ratio variations under different edge conditions.

The natural vibration of parallelogram plates was computed by Gorman [14] under simply supported edge conditions using the superposition approach. Leissa et al. [15] examined the time period of skew orthotropic plates with circular variations in density and thickness under different edge conditions. The primary aim of this study is to demonstrate the impact of parallelogram-shaped plates with thickness and density profiles in two and one dimensions, respectively, along with a linear thermal gradient, at SCCC and CSCC edge conditions. Additionally, the authors conducted a convergence study of the frequency modes for three types of plates: parallelogram, rectangle, and square, under various combinations of edge conditions. To validate the key results obtained, the authors compared them with results already cited in the literature, in terms of the time period and frequency modes of the studied plates.

Method of Solution

A parallelogram (non-homogeneous) shaped plate made having angle of skewness θ, length *a*, breadth *b* is described using skew coordinates ψ=y sec θ, ζ = x-tan θ mentioned in Figures 1 and 2.

The expression for kinetic (T_S) and strain (V_S) energy are given by Leissa [15]:

$$T_s = \frac{1}{2}\omega^2 \cos\theta \int_0^a \int_0^b \rho l \Phi^2 d\zeta d\psi, \tag{1}$$

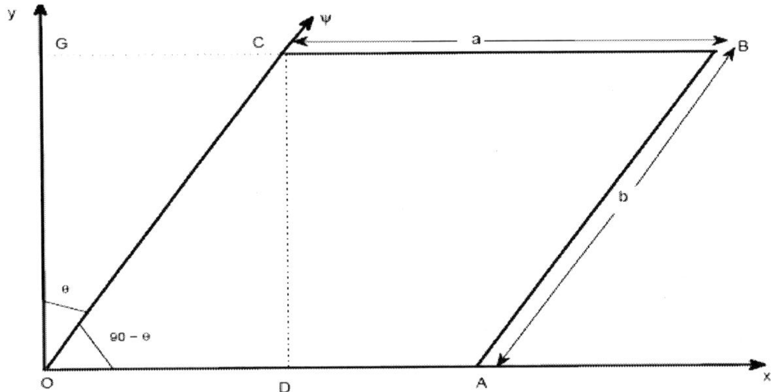

Figure 1. A Parallelogram-shaped plate having θ as the angel of skewness.

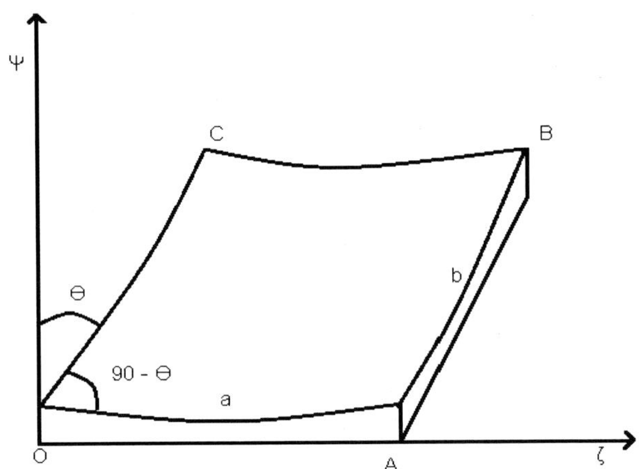

Figure 2. A Parallelogram-shaped plate with a thickness of a circular (two-dimensional) profile.

$$V_s = \frac{1}{2\cos^3\theta} \iint D_1 \begin{bmatrix} \left(\frac{\partial^2 \Phi}{\partial \zeta^2}\right)^2 - 4\sin\theta \left(\frac{\partial^2 \Phi}{\partial \zeta^2}\right)\left(\frac{\partial^2 \Phi}{\partial \zeta \partial \psi}\right) \\ +2\left(\sin^2\theta + \nu\cos^2\theta\right)\left(\frac{\partial^2 \Phi}{\partial \zeta^2}\right)\left(\frac{\partial^2 \Phi}{\partial \psi^2}\right) \\ +2\left(1+\sin^2\theta - \nu\cos^2\theta\right)\left(\frac{\partial^2 \Phi}{\partial \zeta \partial \psi}\right)^2 \\ -4\sin\theta\left(\frac{\theta^2 \Phi}{\partial \zeta \theta \psi}\right)\left(\frac{\partial^2 \Phi}{\partial \psi^2}\right) + \left(\frac{\partial^2 \Phi}{\partial \psi^2}\right)^2 \end{bmatrix} d\zeta\, d\psi \tag{2}$$

where $D_1 = \frac{El^3}{12(1-\nu^2)}$ flexural rigidity, ν: Poisson's ratio, E: Young's modulus and Φ is a deflection function. The Rayleigh-Ritz technique states:

$$J = \delta(V_s - T_s) = 0 \tag{3}$$

On substituting Equations (1)- (2) in Equation (3), we get:

$$J = \frac{1}{2\cos^3\theta}\iint D_1 \begin{bmatrix} \left(\frac{\partial^2\Phi}{\partial\zeta^2}\right)^2 - 4\sin\theta\left(\frac{\partial^2\Phi}{\partial\zeta^2}\right)\left(\frac{\partial^2\Phi}{\partial\zeta\partial\psi}\right) \\ +2\left(\sin^2\theta + \nu\cos^2\theta\right)\left(\frac{\partial^2\Phi}{\partial\zeta^2}\right)\left(\frac{\partial^2\Phi}{\partial\psi^2}\right) \\ +2\left(1+\sin^2\theta - \nu\cos^2\theta\right)\left(\frac{\partial^2\Phi}{\partial\zeta\partial\psi}\right)^2 \\ -4\sin\theta\left(\frac{\theta^2\Phi}{\partial\zeta\theta\psi}\right)\left(\frac{\partial^2\Phi}{\partial\psi^2}\right) + \left(\frac{\partial^2\Phi}{\partial\psi^2}\right)^2 \end{bmatrix} d\zeta\, d\psi$$

$$-\frac{1}{2}\omega^2\cos\theta\iint \rho l \Phi^2 d\zeta\, d\psi = 0 \tag{4}$$

The plate's thickness (l) was taken as circular in two dimensions, whereas the density (ρ) was considered to be circular in one dimension:

$$l = l_0\left[1+\beta_1\left(1-\sqrt{1-\frac{\zeta^2}{a^2}}\right)\right]\left[1+\beta_2\left(1-\sqrt{1-\frac{\psi^2}{b^2}}\right)\right], \rho = \rho_0\left[1-m\left(1-\sqrt{1-\frac{\zeta^2}{a^2}}\right)\right], \tag{5}$$

where $l_0$and ρ_0 represent the plate thickness and plate density at the origin, respectively. Also, the taper parameters, represented by β_1 and β_2, are considered to lie between [0, 1], whereas the non-homogeneity parameter m lies between [0, 1). The thermal gradient (two-dimensional) variations on the plate, considered to be linear, are expressed by Khanna & Arora [16]:

$$\eta = \eta_0\left(1-\frac{\zeta}{a}\right)\left(1-\frac{\psi}{b}\right), \tag{6}$$

where η represents the temperature excess at any point on the plate, and η_0 is the temperature excess at the origin. The modulus of elasticity's dependence on the thermal gradient is taken as:

$$\mathrm{E} = \mathrm{E}_0(1-\gamma\eta) \tag{7}$$

where γ is the slope of variation. And E_0 is the Young's modulus at the mentioned thermal gradient (i.e., $\eta = 0$). Plugging Equation (6) into Equation (7) yields:

$$E = E_0\left[1-\alpha\left(1-\frac{\zeta}{a}\right)\left(1-\frac{\psi}{b}\right)\right], \tag{8}$$

where $\alpha = \gamma\eta_0$, (with $0 \leq \alpha < 1$) represents the thermal gradient. Substituting Equations (5) and (8) into Equation (4), we get:

$$J = \frac{D_0}{2}\int_0^a\int_0^b \left\{ \begin{array}{l} \left[1-\alpha\left\{1-\frac{\zeta}{a}\right\}\left\{1-\frac{\psi}{b}\right\}\right]\left[(1+\beta_1\Lambda_1)(1+\beta_2\Lambda_2)\right]^3 \\ \left[\begin{array}{l} \left(\frac{\partial^2\Phi}{\partial\zeta^2}\right)^2 - 4\Upsilon\sin\theta\left(\frac{\partial^2\Phi}{\partial\zeta^2}\right)\left(\frac{\partial^2\Phi}{\partial\zeta\partial\psi}\right) \\ +2\Upsilon^2\left(\sin^2\theta+\nu\cos^2\theta\right)\left(\frac{\partial^2\Phi}{\partial\zeta^2}\right)\left(\frac{\partial^2\Phi}{\partial\psi^2}\right) \\ +2\Upsilon^2\left(1+\sin^2\theta-\nu\cos^2\theta\right)\left(\frac{\partial^2\Phi}{\partial\zeta\partial\psi}\right)^2 \\ -4\Upsilon^3\sin\theta\left(\frac{\theta^2\Phi}{\partial\zeta\theta\psi}\right)\left(\frac{\partial^2\Phi}{\partial\psi^2}\right)+\Upsilon^4\left(\frac{\partial^2\Phi}{\partial\psi^2}\right)^2 \end{array}\right] \end{array} \right\} d\zeta\, d\psi$$

$$-\lambda^2\cos^4\theta\int_0^a\int_0^b\left[(1-m\Lambda_1)(1+\beta_1\Lambda_1)(1+\beta_2\Lambda_2)\right]^3\Phi^2 d\zeta\, d\psi \tag{9}$$

where $D_0 = E_0 l_0^3/12\left(1-\nu^2\right)$, $\Lambda_1 = \left(1-\sqrt{1-\frac{\zeta^2}{a^2}}\right), \Lambda_2 = \left(1-\sqrt{1-\frac{\psi^2}{b^2}}\right)$, $\lambda^2 = \rho_0\omega^2 l_0 a^4 / D_0$ and $\Upsilon = (a/b)$.

The function represents the deflection of the plate must satisfy all the boundary conditions as shown by Bhardwaj et al. [17]:

$$\Phi(\zeta,\psi)=\left[\left(\frac{\zeta}{a}\right)^{A}\left(\frac{\psi}{b}\right)^{B}\left(1-\frac{\zeta}{a}\right)^{C}\left(1-\frac{\psi}{b}\right)^{D}\right]\times\left[\sum_{i=0}^{N}\Omega_i\left\{\left(\frac{\zeta}{a}\right)\left(\frac{\psi}{b}\right)\left(1-\frac{\zeta}{a}\right)\left(1-\frac{\psi}{b}\right)\right\}^{i}\right], \tag{10}$$

The first part describes the boundary conditions, which change according to values A, B, C, and D. Values of 2,1 and 0 denote clamped, supported, and free boundary conditions, respectively. The second part indicates the number of frequency modes, while Ω_i, i= 0, 1, 2, N denotes arbitrary constants. For minimizing the functional mentioned in Equation(9), the following constraints have to be satisfied:

$$\frac{\delta J}{\partial \Omega_i}=0, \quad i=0,1,2,3...N. \tag{11}$$

On solving Equation (11), we obtain a system of the homogeneous equations in Ω_i. The non-zero Equation gives the desired frequency equation as:

$$|P-\lambda^2 Q|=0, \tag{12}$$

where $P=\left[p_{ij}\right]_{N+1}$ and $Q=\left[q_{ij}\right]_{N+1}$ are square matrix of order $(n+1), i=0,1,2...N$ and $j=0,1,2...N$. Equation(13) mentioned below is used for computing the time period for frequency (λ), obtained from Equation (12):

$$K=\frac{2\pi}{\lambda}, \tag{13}$$

Numerical Work and Discussion

In this segment, the authors report the results obtained for parallelogram-shaped plates with thickness and density profiles in two and one dimensions, respectively, along with a linear thermal gradient. The parameters that were kept fixed in this study were aspect ratio a/b=1.5, angle of skewness θ =30°, and Poisson's ratio ν=0.345. The results are divided into different scenarios based on the edge conditions SCCC and CSCC. The various scenarios investigated are listed below:

- Case I: In the first set of investigations, we varied the β_1 *and* β_2 for fixed values of α and$m$.
- Case II: In the second scenario, for a fixed value of m, we varied α and β_1, β_2.
- Case III: In the last scenario, we varied m and β_1 *and* β_2 for a fixed value of m.

Now, we will summarize the results obtained for each of the scenarios separately in tabulated form (Refer Tables 1-3).

In Case I, we varied the β_1 and β_2 for a fixed value of α and m. Table 1 represents the period K of the parallelogram under all the aforementioned boundary conditions with respect to β_1, β_2 and for $m = 0.6$, $\alpha = 0.4$. The tapering parameters have been varied from 0.0 to 1.0. The conclusions mentioned below are from Table 1.

- For increasing values of β_1 and β_2, the time period$K$ decreases at each edge condition.
- The time period K obtained was highest at *SCCC* and lowest at *CSCC* edge condition.
- Decremental rate with respect to β_1 was highest at the *CSCC* edge condition and lowest at the *SCCC* edge condition.
- The decremental rate with respect to β_2 was highest at the *SCCC* edge condition and lowest at the *CSCC* edge condition.

Now, we switch our attention to Case II, where we varied$\alpha$and β_1, β_2 forms. Table 2 highlights how the period K of a parallelogram plate varies with α for SCCC and CSCC edge conditions, with $m=0.6$ and β_1,β_2 ranging from 0.0 to 1.0. The facts observed in Table 2 are as follows:

- The time period K increased as the α values increased, whereas they decreased for increasing β_1 and β_2 values.
- The value of time period K was highest at *SCCC* and lowest at *CSCC* edge condition with respect to α.
- The incremental rate with respect to α was highest at the *CSCC* boundary condition and lowest at the *SCCC* boundary condition.

Table 1. Time period K of a parallelogram plate varies with β_1, β_2, for SCCC and CSCC edge conditions

	$m = 0.6, \alpha = 0.4$												
	β_1	$\beta_2 = 0.0$		$\beta_2 = 0.2$		$\beta_2 = 0.4$		$\beta_2 = 0.6$		$\beta_2 = 0.8$		$\beta_2 = 1.0$	
		k_1	k_2	k_1	k_2	k_1	k_2	k_1	k_2	k_1	k_2	k_1	k_2
SCCC	0.0	0.02809	0.08807	0.02642	0.08298	0.02478	0.07807	0.02322	0.07345	0.02176	0.06914	0.02039	0.06517
	0.2	0.02754	0.08640	0.02590	0.08141	0.02430	0.07659	0.02278	0.07206	0.02134	0.06784	0.02001	0.06395
	0.4	0.02699	0.08471	0.02539	0.07982	0.02383	0.07511	0.02233	0.07067	0.02092	0.06653	0.01962	0.06272
	0.6	0.02644	0.08302	0.02488	0.07824	0.02335	0.07363	0.02189	0.06928	0.02051	0.06523	0.01923	0.06150
	0.8	0.02589	0.08133	0.02437	0.07666	0.02287	0.07215	0.02144	0.06789	0.02010	0.06393	0.01884	0.06028
	1.0	0.02535	0.07966	0.02386	0.07509	0.02240	0.07068	0.02100	0.06652	0.01969	0.06265	0.01846	0.05908
CSCC	0.0	0.01517	0.10735	0.01514	0.10397	0.01510	0.10044	0.01504	0.09682	0.01497	0.09316	0.01488	0.08951
	0.2	0.01472	0.10423	0.01469	0.10096	0.01465	0.09755	0.01460	0.09405	0.01453	0.09050	0.01444	0.08697
	0.4	0.01429	0.10115	0.01426	0.09799	0.01422	0.09470	0.01417	0.09131	0.01410	0.08788	0.01402	0.08447
	0.6	0.01387	0.09812	0.01384	0.09508	0.01381	0.09189	0.01376	0.08862	0.01369	0.08531	0.01361	0.08200
	0.8	0.01346	0.09517	0.01344	0.09222	0.01341	0.08914	0.01336	0.08598	0.01329	0.08279	0.01321	0.07959
	1.0	0.01307	0.09228	0.01305	0.08944	0.01302	0.08647	0.01297	0.08342	0.01291	0.08033	0.01283	0.07724

Table 2. Time period K of a parallelogram plate varies with α for SCCC and CSCC edge conditions

	$m = 0.6$												
	α	$\beta_1 = \beta_2 = 0.0$		$\beta_1 = \beta_2 = 0.2$		$\beta_1 = \beta_2 = 0.4$		$\beta_1 = \beta_2 = 0.6$		$\beta_1 = \beta_2 = 0.8$		$\beta_1 = \beta_2 = 1.0$	
		k_1	k_2	k_1	k_2	k_1	k_2	k_1	k_2	k_1	k_2	k_1	k_2
SCCC	0.0	0.02621	0.08224	0.02434	0.07656	0.02253	0.07108	0.02082	0.06591	0.01922	0.06110	0.01773	0.05666
	0.2	0.02710	0.08500	0.02508	0.07887	0.02315	0.07301	0.02133	0.06753	0.01964	0.06247	0.01809	0.05783
	0.4	0.02809	0.08807	0.02590	0.08141	0.02383	0.07511	0.02189	0.06928	0.02010	0.06393	0.01846	0.05908
	0.6	0.02919	0.09151	0.02680	0.08420	0.02456	0.07740	0.02249	0.07117	0.02059	0.06551	0.01886	0.06041
	0.8	0.03044	0.09539	0.02781	0.08731	0.02537	0.07991	0.02314	0.07322	0.02111	0.06721	0.01929	0.06183
CSCC	0.0	0.01369	0.09995	0.01330	0.09443	0.01291	0.08896	0.01252	0.08360	0.01213	0.07843	0.01174	0.07346
	0.2	0.01517	0.10735	0.01395	0.09752	0.01352	0.09169	0.01309	0.08600	0.01267	0.08051	0.01225	0.07528
	0.4	0.01517	0.10735	0.01469	0.10096	0.01422	0.09470	0.01376	0.08862	0.01329	0.08279	0.01283	0.07724
	0.6	0.01610	0.11175	0.01558	0.10482	0.01505	0.09804	0.01453	0.09151	0.01401	0.08527	0.01350	0.07938
	0.8	0.01724	0.11679	0.01664	0.10918	0.01604	0.10180	0.01545	0.09472	0.01487	0.08802	0.01429	0.08171

Table 3. Time period K of a parallelogram plate varies with m for SCCC and CSCC edge conditions

	$\alpha = 0.6$												
	m	$\beta_1 = \beta_2 = 0.0$		$\beta_1 = \beta_2 = 0.2$		$\beta_1 = \beta_2 = 0.4$		$\beta_1 = \beta_2 = 0.6$		$\beta_1 = \beta_2 = 0.8$		$\beta_1 = \beta_2 = 1.0$	
		k_1	k_2	k_1	k_2	k_1	k_2	k_1	k_2	k_1	k_2	k_1	k_2
SCCC	0.0	0.02976	0.09415	0.02733	0.08668	0.02505	0.07972	0.02293	0.07334	0.02100	0.06755	0.01924	0.06232
	0.2	0.02958	0.09328	0.02716	0.08586	0.02489	0.07896	0.02278	0.07262	0.02086	0.06688	0.01912	0.06169
	0.4	0.02939	0.09240	0.02698	0.08504	0.02472	0.07818	0.02264	0.07190	0.02073	0.06620	0.01899	0.06105
	0.6	0.02919	0.09151	0.02680	0.08420	0.02456	0.07740	0.02249	0.07117	0.02059	0.06551	0.01886	0.06041
	0.8	0.02900	0.09061	0.02662	0.08336	0.02439	0.07661	0.02233	0.07043	0.02045	0.06482	0.01873	0.05976
CSCC	0.0	0.01694	0.11716	0.01640	0.10995	0.01587	0.10289	0.01533	0.09608	0.014800	0.08958	0.01427	0.08342
	0.2	0.01667	0.11538	0.01613	0.10826	0.01560	0.10130	0.01507	0.09458	0.01454	0.08816	0.01402	0.08209
	0.4	0.01639	0.11358	0.01586	0.10655	0.01533	0.09968	0.01480	0.09306	0.01428	0.08673	0.01376	0.08075
	0.6	0.01610	0.11175	0.01558	0.10482	0.01505	0.09804	0.01453	0.09151	0.01401	0.08527	0.01350	0.07938
	0.8	0.01581	0.10989	0.01529	0.10305	0.01477	0.09637	0.01425	0.08993	0.01374	0.08379	0.01324	0.07799

- Decremental rate with respect to both β_1, β_2 was highest at *SCCC* edge condition and lowest at *CSCC* boundary condition.

Ultimately, we discussed the results obtained in Case III, where we varied m and β_1, β_2 for α.

Table 3 highlights how the time period K of a parallelogram plate varies with m for SCCC and CSCC edge conditions, with α =0.6 and β_1, β_2 ranging from 0.0 to 1.0. Table 3 reveals that:

- As the value of m along with both β_1 *and* β_2 were increased, the time period$K$ decreased for both the chosen boundary conditions.
- The decremental rate with respect to m was highest at the *CSSC* edge condition and lowest at the *SCCC* edge condition.
- Decremental rate with respect to both β_1, β_2 was highest at *SCCC* edge condition and lowest at *CSCC* edge condition.

Results Comparison

In this segment, we made comparisons with the following already cited results in the literature in terms of the time period and frequency modes of the studied plate:

- Frequency modes λ of the concerned plate achieved in Arora, [18] at CCCC boundary condition concerning the skewness angle θ (see Refer Table 4).
- The time period K of the concerned plate achieved by Gupta et al., [19] with respect to β_1 mentioned in Table 5 for CCCC boundary conditions.
- Modes of frequency λ of the concerned plate achieved in Sharma et al. [20] concerning m mentioned in Table 6 for CCCC boundary condition.

The results are shown in tabulated form. The comparative analysis presents the objective of the present study well.

Table 4. Comparative analysis of modes of frequency corresponding to θ for a/b = 1.5∘

		$\alpha = m = 0.0$											
	θ	$\beta_1 = 0.0$		$\beta_1 = 0.2$		$\beta_1 = 0.4$		$\beta_1 = 0.6$		$\beta_1 = 0.8$		$\beta_1 = 1.0$	
		λ_1	λ_2	λ_1	λ_2	λ_1	λ_2	λ_1	λ_2	λ_1	λ_2	λ_1	λ_2
CCCC	0°	54.15	217.54	55.93	223.98	57.78	230.70	59.70	237.66	61.68	244.86	63.72	252.27
		60.84	***240.47***	***69.07***	***273.63***	***77.53***	***307.67***	***86.16***	***342.31***	***94.91***	***377.38***	***103.74***	***412.76***
	30°	74.00	294.60	76.45	303.38	79.01	312.54	81.68	322.04	84.43	331.87	87.27	342.00
		84.61	***329.83***	***96.01***	***375.28***	***107.77***	***421.95***	***119.76***	***469.45***	***131.93***	***517.55***	***144.22***	***566.09***
	60°	232.24	910.49	240.10	937.95	248.34	966.62	256.92	996.42	265.81	1027.28	275.00	1059.10
		273.10	***1042.66***	***309.74***	***1186.16***	***347.61***	***1333.59***	***386.31***	***1483.69***	***425.60***	***1635.73***	***465.31***	***1789.15***

In Table 4, a comparison study of frequency modes λ attained by present research and in literature by Arora [18]) for the same set of chosen boundary conditions was done with respect to θ and fixed values of $m = 0.0$, $\alpha = 0.0$ and variable value of β_1, β_2 from 0.0 to 1.0. Here, the authors exclude the tapering parameter β_2 because this plate parameter is not considered in published results by Arora [18]. The comparison concluded that for increasing θ, frequency modes attained in the current study are lesser than what has been obtained in the published results by Arora [18] due to circular variation in β_1.

Italicized and bold values are obtained from the published results of Arora [18].

In Table 5, a comparison study of time period K attained by present research and in published results by Gupta et al. [19] for the same set of chosen boundary conditions were done with respect to β_1 and fixed values of $m = 0.0$, $\alpha = 0.0$ and for two different value of skew angle θ (i.e., $\theta = 0^\circ$ and 45°). The comparison concluded that for increasing β_1, the frequency modes attained in the current study are lesser than what has been obtained in the literature by Gupta et al. [19].

Table 5. Comparative analysis of time period of corresponding to to β_1 for a/b =1.5

		$\theta = 0^\circ$		$\theta = 45^\circ$	
	β_1	k_1	k_2	k_1	k_2
CCCC	0.0	0.02888	0.11602	0.01400	0.05530
		0.03635	***0.14264***	***0.02397***	***0.09030***
	0.2	0.02805	0.11234	0.01359	0.05351
		0.03366	***0.13178***	***0.02219***	***0.08463***
	0.4	0.02723	0.10874	0.01319	0.05175
		0.03066	***0.12182***	***0.02037***	***0.07802***
	0.6	0.02643	0.10524	0.01280	0.05004
		0.02806	***0.11295***	***0.01846***	***0.07214***
	0.8	0.02566	0.10186	0.01242	0.0439
		0.02589	***0.10406***	***0.06714***	***0.01709***

Italicized and bold values are obtained from the published results of Gupta et al. [19]. In Table 6, a comparison study of frequency modes λ attained by present research and in available published results by Sharma et al. [20] for the same set of chosen boundary conditions was done with respect to m and fixed values of $\alpha = 0.0$ and variable value of β_1, β_2 from 0.0 to 0.8.

Table 6. Comparative analysis of modes of frequency λ corresponding to *m* for a/b = 1.5 and θ = 30◦

		$\alpha = 0.0$									
		$\beta_1 = \beta_2 = 0.0$		$\beta_1 = \beta_2 = 0.2$		$\beta_1 = \beta_2 = 0.4$		$\beta_1 = \beta_2 = 0.6$		$\beta_1 = \beta_2 = 0.8$	
	m	λ_1	λ_2	λ_1	λ_2	λ_1	λ_2	λ_1	λ_2	λ_1	λ_2
CCCC	0.0	74.00	294.60	80.84	321.15	88.39	351.12	96.61	384.50	105.45	421.24
		74.00	***294.60***	***99.20***	***396.12***	***128.55***	***514.68***	***162.06***	***650.23***	***199.75***	***802.77***
	0.2	75.15	299.82	82.10	326.98	89.79	357.64	98.15	391.78	107.14	429.38
		78.00	***310.53***	***104.54***	***417.32***	***135.44***	***541.98***	***170.72***	***684.50***	***210.39***	***844.86***
	0.4	76.35	305.33	83.43	333.14	91.25	364.54	99.76	399.51	108.91	438.02
		82.73	***329.37***	***110.84***	***442.32***	***143.57***	***574.15***	***180.94***	***724.84***	***222.95***	***894.34***
	0.6	77.62	311.16	84.83	339.67	92.79	371.86	101.45	407.72	110.77	447.21
		88.44	*352.11*	*118.45*	*472.43*	*153.37*	*612.83*	*193.24*	*773.26*	*238.06*	*953.69*
	0.8	78.94	317.35	86.30	346.61	94.41	379.66	103.24	416.47	112.73	457.02
		95.53	***380.33***	***127.87***	***509.67***	***165.50***	***660.55***	***208.45***	***832.89***	***256.73***	***1026.60***

The comparison concluded that for increasing m, the time period attained in the current study is lesser than what has been obtained in the published results by Sharma et al. [20]. Also, no, te that the value of the time period exactly matched when the values of β_1, β_2 and m become 0.0.

Italicized and bold values are obtained from the published results of Sharma et al. [20].

Conclusion

In this study, an analysis of the time period K of a parallelogram-shaped plate for SCCC, CSCC, SSCC, CSSC, SCSS, SSCS, and SSSC edge conditions corresponding to various values of plate parameters is presented. The authors would like to draw the following conclusions based on the data obtained from numerical computations and comparisons:

- The frequency modes (λ) for a circular profile in thickness (present research) were lower in magnitude than the frequency modes for a sinusoidal profile in thickness by Arora [18] for the CCCC edge condition (see Table 4).
- The time period for a circular profile in thickness (present research) was lower in magnitude than the time period for a parabolic profile in thickness by Gupta et al. [19] for the CCCC edge condition (see Table 5).
- The frequency modes (λ) for circular variation in density (current study) were lower in magnitude than those for linear variation in density by Sharma et al. [20] for CCCC and CSCS boundary conditions (see Table 6).

Disclaimer

None.

References

[1] Pagani A., Azzara R. & Carrera E. (2023). Geometrically nonlinear analysis and vibration of in-plane-loaded variable angle tow composite plates and shells, *Acta Mechanica,* 234(1), 85–108.

[2] Vinh V. P. (2023). Deflections, stresses and free vibration analysis of bi-functionally graded sandwich plates resting on Pasternak elastic foundations via a hybrid quasi-3D theory, *Mechanics Based Design of Structures and Machines*,51(4), 2323–2354.

[3] Farag A. & Ashour A. (2000). Free vibration of orthotropic skew plates, *J. Vib. Acoust.*, 122(3), 313–317.

[4] Sharma A. & Sharma A. K. (2016). Mathematical modelling of vibration on parallelogram plate with non-homogeneity y effect, *Romanian Journal of Acoustics and Vibration,*13(1), 53-57.

[5] Sharma A. K. & Dhiman M. K. (2019). The vibration of a non-homogeneous parallelogram plate (SSSS) has a conflicting bi-illustrative thickness and bi-direct temperature variation, *International Journal of Applied Engineering Research*, 14(2), 516–522.

[6] Gupta A. K., Kumar A. & Gupta D. V. (2012). Vibration of visco-elastic orthotropic parallelogram plate with parabolically thickness variation, *Annals of the Faculty of Engineering Hunedoara*, 10(2), 61–70.

[7] Gupta A., Kumar A. & Gupta D. (2007). Vibration of visco-elastic orthotropic parallelogram plate with linear thickness variation, *International Conference in World Congress on Engineering and Computer Science*, 800–803.

[8] Sharma S. K. & Sharma A. K. (2015). Mathematically study the vibration of a visco-elastic parallelogram plate, *Mathematical Models in Engineering*,1(1), 12–19.

[9] Kumar A., Lather N., Bhardwaj R., Mani N. & Sharma A. (2018). Effect of linear variation in density and circular variation in Poisson ratio on the time period of vibration of a rectangular plate, *Vibroengineering Procedia,* 21, 14–19.

[10] Conway H. & Farnham K. (1965). The free flexural vibrations of triangular, rhombic and parallelogram plates and some analogies, *International Journal of Mechanical Sciences*, 7(12), 811–816.

[11] Lather N. & Sharma A. (2022). The behaviour of frequencies of an orthotropic rectangular plate with circular variations in thickness and density. *VibroengineeringProcedia*, 41, 71–76.

[12] Lather N. & Sharma A. (2019). Natural vibration of the skew plate on a different set of boundary conditions with a temperature gradient, *Vibroengineering Procedia*, 22, 74–80.

[13] Kumar A., Lather N. & Sharma A. (2019). Analysis of the time period of the isotropic square plate on clamped and simply supported conditions, *AIP Conference Proceedings*, 2142.

[14] Gorman D. (1991). Accurate analytical-type solutions for the free vibration of simply-supported parallelogram plates. *Journal of Applied Mechanics,* 58(1), 203-208.

[15] Leissa A. W. (1969). Vibration of plates, *Scientific and Technical Information Division, National Aeronautics and Space Administration*, 160.

[16] Khanna A. & Arora P. (2013). Effect of sinusoidal thickness variation on vibrations of non-homogeneous parallelogram plate with bi-linearly temperature variations, *Indian Journal of Science and Technology,* 6(9), 5228–5234.

[17] Bhardwaj R., Mani N. & Sharma A. (2021). The time period of transverse vibration of the skew plate with parabolic temperature variation, *Journal of Vibration and Control*, 27(3-4), 323-331.

[18] Arora P. (2015). Some mathematical models on the vibration of parallelogram plate of variable thickness-effect of two-dimensional temperature variations, *PhD Thesis, Maharishi Markandeshwar University Mullana, Ambala (Haryana) India.*

[19] Gupta A. K., Kumar A., Gupta Y. K., et al. (2010). Vibration of visco-elastic parallelogram plate with parabolic thickness variation, *Applied Mathematics,* 1(2), 128–136.

[20] Sharma A., Bhardwaj R., Lather N., Ghosh S., Mani N. & Kumar K. (2022). Time period of thermal-induced vibration of the skew plate with two-dimensional circular thickness, *Mathematical Problems in Engineering*, Article ID 8368194.

Chapter 5

Unraveling Thermal Stress in Rotating Disks: An Analytical Approach for Transversely Isotropic Materials

Pankaj Thakur, PhD
Priya Gulial*, MSc
and Palvinder Thakur, MSc
Faculty of Science and Technology, ICFAI University Himachal Pradesh, Himachal Pradesh, India

Abstract

This study develops an analytical framework to examine thermal stress in rotating disks of transversely isotropic materials. By solving governing equations, it reveals the impact of anisotropy on stress distribution under rotational and thermal loads. The findings enhance understanding of material behavior, aiding in the design of thermally resilient engineering systems.

Keywords: thermal stress, rotating disks, stress distribution, transversely isotropic materials, analytical approach

* Corresponding Author's Email: gulialpriya184@gmail.com

In: Thermal Modeling Reimagined
Editors: Pankaj Thakur and Jatinder Kaur
ISBN: 979-8-89530-463-1

Introduction

Rotating disks are vital in numerous engineering applications, serving as essential components in turbines, engines, and other rotating machinery. However, thermal stresses during operation can significantly impact their performance and structural integrity. Therefore, engineers must conduct thermal stress analysis to ensure the reliability and longevity of these critical components. A key challenge in analyzing thermal stress in rotating disks arises from the diverse mechanical properties of the materials used, particularly those with transversely isotropic characteristics. Transversely isotropic materials exhibit varying mechanical behavior along different axes, making stress analysis more complex than for isotropic materials.

As a result, engineers require a specialized analytical approach to predict thermal stress distributions in such disks accurately. This study develops an analytical framework tailored specifically for transversely isotropic materials to clarify the complex interplay between thermal effects and material properties in rotating disks. By addressing this gap in existing research, the study provides engineers and designers with a robust tool for predicting and mitigating thermal stresses in rotating disk systems. The significance of this research spans industries such as aerospace, automotive, and power generation, where rotating machinery is widespread. Enhanced understanding and predictive capabilities regarding thermal stress in rotating disks can improve component design, performance optimization, and overall system reliability.

Furthermore, by offering insights into the behavior of transversely isotropic materials under thermal loading conditions, this study contributes to advancements in materials science and engineering. In this introduction, we outline the study's motivation, objectives, and scope, preparing the reader for the subsequent sections, which detail the analytical methodology, present results, and discuss implications for engineering practice and future research. Through this interdisciplinary approach, we aim to advance knowledge and promote innovation in the analysis and design of rotating disk systems subjected to thermal stresses.

Various references extensively document the analytical exploration of elastic-plastic rotating disks made from transversely isotropic materials [1-4]. The use of rotating disks in engineering and scientific fields has generated considerable interest in creep phenomena, maintaining its status as a vibrant research domain. Swainger [5] investigated the deformation behavior of these disks, while Ghose [6] focused on the thermal effects influencing the transverse vibrations of spinning disks with varying thicknesses. Murakami & Konishi [7] established a constitutive equation tailored for transversely isotropic materials. Furthermore, Thakur et al. [8] employed Seth's transition theory to analyze thermal creep transition stresses in thin rotating disks with a shaft and variable density. Gupta & Thakur [9] also examined thermo-elastic plastic transitions in thin rotating disks with inclusions, applying transition theory to their studies.

This chapter provides an analytical framework for thermal stress in rotating disks of transversely isotropic materials, highlighting fundamental equations and transition points. It examines the effects of thermal expansion and rotation, offering new insights compared to previous studies and emphasizing practical implications for engineering applications.

Material

Beryl, a cyclosilicate, embodies transverse isotropy in its structure. Characterized by a lustrous sheen, it commonly occurs in hexagonal prism formations, culminating in pinacoid terminations. Brass, on the other hand, is an isotropic alloy of copper and zinc. Renowned for its electrical and thermal conductivity, strength, hardness, and malleability, it is a versatile material used in various applications.

Basics Governing Equations

For analysis purposes, we consider a homogeneous rotating disk mounted on a rigid shaft made of a transversely isotropic material with constant density. The disk features a central bore with an inner radius a and an outer radius b ($b>a$). We apply a temperature Θ to the inner surface of the disk's central bore, assuming the rotating disk is fixed to the shaft (see Figure 1). Therefore, we define the boundary conditions for the problem as follows:

$$u = 0,\ r = a\ ,\ \tau_{rr} = 0\ ,\ r = b \tag{1}$$

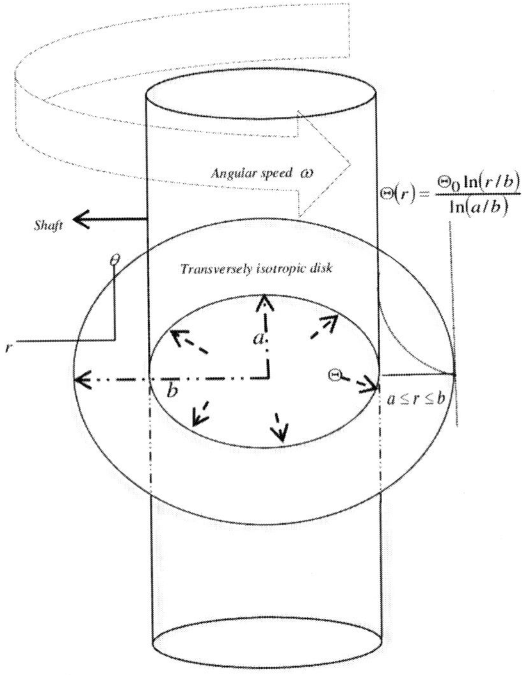

Figure 1. Geometry of disk.

The generalized components of strain are given by Seth [10]:

$$e_{rr} = \frac{1}{n}\left[-(\beta + r\beta')^n + 1\right],\ e_{\theta\theta} = \frac{1}{n}\left[-\beta^n + 1\right],\ e_{zz} = \frac{1}{n}\left[-(1-d)^n + 1\right],$$
$$e_{r\theta} = 0 = e_{\theta z} = e_{zr} \tag{2}$$

where β a position function depends only on x and y. References [2, 10] provide the thermo-elastic constitutive equations:

$$\tau_{rr} = c_{11}\varepsilon_{rr} + (c_{11} - 2c_{66})\varepsilon_{\theta\theta} + c_{13}\varepsilon_{zz} - \beta_1\Theta$$
$$\tau_{\theta\theta} = (c_{11} - 2c_{66})\varepsilon_{rr} + c_{11}\varepsilon_{\theta\theta} + c_{13}\varepsilon_{zz} - \beta_1\Theta$$
$$\tau_{r\theta} = \tau_{\theta z} = \tau_{zr} = \tau_{zz} = 0 \tag{3}$$

where β_1 α_1 represents the coefficient of linear thermal expansion along the axis of symmetry and α_2 corresponds to the quantities orthogonal to the axis of symmetry. The parameters c_{ij} are the elastic material constants and Θ denote the temperature change. By substituting Equation (2) into Equation (3), we derive the stresses, yielding:

$$\tau_{rr} = \frac{c_{11}}{n}\left[-(\beta + r\beta')^n + 1\right] + (c_{11} - 2c_{66})\frac{1}{n}\left[-\beta^n + 1\right] + c_{13}e_{zz} - \beta_1\Theta \tag{4}$$

$$\tau_{\theta\theta} = \frac{(c_{11} - 2c_{66})}{n}\left[-(\beta + r\beta')^n + 1\right] + \frac{c_{11}}{n}\left[-\beta^n + 1\right] + c_{13}e_{zz} - \beta_1\Theta \tag{5}$$

$$\tau_{r\theta} = \tau_{\theta z} = \tau_{zr} = \tau_{zz} = 0$$

The temperature field that satisfies the heat equation is given in reference [2] as follows: $\Theta = \overline{\Theta}_0 \ln(r/b)$ where

$$\overline{\Theta}_0 = \frac{\Theta_0}{\ln(a/b)} \tag{6}$$

The Equation of equilibrium for stress-strain is given as:

$$\frac{d}{dr}(\tau_{rr}) + \frac{(\tau_{rr} - \tau_{\theta\theta})}{r} + \rho\omega^2 r = 0 \tag{7}$$

Asymptotic solution at the transition points: By substituting Equations (4) and (5) into Equation (7), the resulting non-linear differential Equation is obtained as

$$c_{11}n\beta^{n+1}P(1+P)^{n-1}\frac{dP}{d\beta} = \begin{bmatrix} -c_{11}n\beta^n P(1+P)^n - (c_{11} - 2c_{66})n\beta^n P \\ +2c_{66}\left[1 - (1+P)^n\right]\beta^n - n\beta_1\overline{\Theta}_0 + n\rho\omega^2 r^2 \end{bmatrix} \tag{8}$$

where $r\beta' = P\beta$ and $\beta' = d\beta/dr$. The transition points β are $P = -1$ (plastic to creep state) and $P \to \pm\infty$ (elastic to plastic state).

Solution

Thakur et al. [10-19] demonstrated the asymptotic solution that leads to the creep state at the transition point through the principal stress differences. The transition function is:

$$T_{trans} = \tau_{rr} - \tau_{\theta\theta} \equiv \frac{2}{n} c_{66}\beta^n \left[1-(1+P)^n\right] \tag{9}$$

By performing logarithmic differentiation of Equation (9) with respect to r, and then substituting Equation (8) into Equation (9) while considering the asymptotic value $P \to -1$, we obtain:

$$\frac{d}{dr}(\ln T_{trans}) = \frac{1}{r}\left\{-2n - \frac{n\rho\omega^2 r^2}{c_{11}\beta^n} + \frac{n}{\beta^n c_{11}}\beta_1\overline{\Theta}_0 + c_1(n-1)\right\} \tag{10}$$

where $c_1 = 2c_{66}/c_{11}$. Integrating Equation (10), we get:

$$T_{trans} = \tau_{rr} - \tau_{\theta\theta} \equiv A_1 r^{-2n+c_1(n-1)} \exp(f) \tag{11}$$

where $f = \frac{n\beta_1\overline{\Theta}_0}{c_{11}}\int \frac{1}{\beta^n r}dr - \frac{n\rho\omega^2}{c_{11}}\int \frac{r}{\beta^n}dr$ and A_1 is a constant of integration. By substituting the expression from Equation (7) into Equation (11), we obtain:

$$\tau_{rr} = -A_1 \int r^{-1-2n+c_1(n-1)} \exp(f) dr - \frac{\rho\omega^2 r^2}{2} + A_2 \tag{12}$$

Imposing the boundary condition $\tau_{rr} = 0$ at $r = b$ in Equation (12), we obtain:

$$A_2 = A_1 \int_{r=b} r^{-1-2n+c_1(n-1)} \exp(f) dr - \frac{\rho\omega^2 b^2}{2} \tag{13}$$

Incorporating Equation (13) into Equation (12) results in:

$$\tau_{rr} = A_1 \int_r^b r^{-1-2n+c_1(n-1)} \exp(f) dr - \frac{\rho\omega^2 (r^2 - b^2)}{2} \tag{14}$$

Applying the expression from Equation (14) within Equation (11) gives us:

$$\tau_{\theta\theta} = \left[A_1 \left[\int_r^b r^{-1-2n+c_1(n-1)} \exp(f) dr - r^{-2n+c_1(n-1)} \exp(f) \right] - \frac{\rho\omega^2 (r^2 - b^2)}{2} \right] \tag{15}$$

The displacement can be determined by utilizing both Equation (9) and Equation (11):

$$\frac{2}{n} c_{66} \eta^n \left[1 - (1+T)^n\right] = A_1 r^{-2n+c_1(n-1)} \exp(f) \tag{16}$$

As $T \to -1$, Equation (16), we get:

$$\eta = \left[\frac{n}{2c_{66}} A_1 r^{-2n+c_1(n-1)} \exp(f) \right]^{\frac{1}{n}} \tag{17}$$

Incorporating Equation (17) within Equation (1), the displacement component results in:

$$u = r - r \left[\frac{n}{2c_{66}} A_1 r^{-2n+c_1(n-1)} \exp(f) \right]^{\frac{1}{n}} \tag{18}$$

Consequently, by enforcing the boundary condition $u = 0$ at r = a in Equation (26), the constant A_1 is found as:

$$A_1 = \frac{2c_{66} a^{2n-c_1(n-1)}}{n \exp(f)|_{r=a}} \tag{19}$$

Incorporating the value of A1 into Equations (14), (15), and (18) results in:

$$\tau_{rr} = \frac{2c_{66}a^{2n-c_1(n-1)}}{n\exp(f)|r=a} \int_r^b r^{-1-2n+c_1(n-1)} \exp(f)dr - \frac{\rho\omega^2(r^2-b^2)}{2} \tag{20}$$

$$\tau_{\theta\theta} = \left\{ \begin{array}{l} \frac{2c_{66}a^{2n-c_1(n-1)}}{n\exp(f)|r=a} \left[\int_r^b r^{-1-2n+c_1(n-1)} \exp(f)dr - r^{-2n+c_1(n-1)} \exp(f) \right] \\ -\frac{\rho\omega^2(r^2-b^2)}{2} \end{array} \right\} \tag{21}$$

$$u = r - r\left[\left(\frac{r}{a}\right)^{-2n+c_1(n-1)} \frac{\exp(f)}{\exp(f)|r=a} \right]^{\frac{1}{n}} \tag{22}$$

Equations (20) - (22) in non-dimensional form:

$$T_r = \frac{2R_0^{2n-c_1(n-1)}}{n\exp(f)|R=R_0} \int_R^1 R^{-1-2n+c_1(n-1)} \exp(f)dR - \frac{\Omega^2(R^2-1)}{2} \tag{23}$$

$$T_\theta = \frac{2R_0^{2n-c_1(n-1)}}{n\exp(f)|R=R_0} \left[\begin{array}{l} \int_R^1 R^{-1-2n+c_1(n-1)} \exp(f)dR \\ - R^{-2n+c_1(n-1)} \exp(f) \end{array} \right] - \frac{\Omega^2(R^2-1)}{2} \tag{24}$$

$$U = R - R\left[\left(\frac{R}{R_0}\right)^{-2n+c_1(n-1)} \frac{\exp(f)}{\exp(f)|R=R_0} \right]^{\frac{1}{n}} \tag{25}$$

where $f = \frac{\beta_4 c_1}{2}\left(\frac{r_0}{M}\right)^n R^n - \frac{n\Omega^2 c_1}{2(n+2)}\left(\frac{r_0}{n}\right)^n R^{n+2} M$ is constant and non-dimensional quantities are introduced as: $R = r/b$ $R_0 = a/b$, $U = u/b$, $\Omega^2 = \rho\omega^2 b^2 / c_{66}$ $\beta_4 = \beta_1\overline{\Theta}_0 / c_{66}$ and.

Equations (23) through (25) match those provided by Thakur & Sethi [12], provided that thermal conditions are ignored.

Results and Discussion

This study provides an in-depth analytical framework for assessing thermal stresses in rotating disks made from transversely isotropic materials, focusing on the complex interactions between thermal expansion and boundary conditions. The analytical approach derives fundamental equations that describe material behavior under rotational and thermal influences. By applying non-dimensional forms to the governing equations, the research offers a clearer understanding of stress distribution patterns and critical transition points, enabling a deeper insight into how these materials respond under varying load conditions. When thermal effects are disregarded, the results closely align with previous findings by Thakur & Sethi [12], reinforcing the validity of the derived model in purely rotational scenarios. These results are significant for designing and optimizing engineering components that operate under both thermal and rotational stresses, providing a foundation for safer and more efficient structural applications across various industrial sectors.

Conclusion

The results of this chapter are as follows:

- This study presents a comprehensive analytical approach to understanding thermal stress in rotating disks made from transversely isotropic materials
- Disregarding thermal conditions, the results align with those presented by Thakur & Sethi [12].
- The findings provide valuable insights for designing and analyzing engineering components exposed to significant thermal and rotational stresses.

Disclaimer

None.

References

[1] Timoshenko, S. P. & Goodier, J. N. (1970). *Theory of Elasticity*, Mc Graw-Hill, New York, USA.

[2] Parkus H. (1976). *Thermo-elasticity*, Vienna: Spriger-verlag.

[3] Johnson W. & Mellor P. B. (1978). *Engineering Plasticity*, London: Von Nostrand Reinhold.

[4] Altenbach H. & Skrzypek J. J. (1999). Creep and damage in materials and structures, Springer Verlag, Berlin.

[5] Swainger K. H. (1956). *Analysis of Deformation*, Chapman & Hall, London; Macmillan, USA, Vol. III, Fluidity,67-68.

[6] Ghose N. C. (1975). Thermal effect on the transverse vibration of spinning disk of variable thickness, *J. Appl. Mech.*, 42, 358-362.

[7] Murakami S. & Konishi K. (1982). An elastic-plastic constitutive equation for transversely isotropic materials and its application to the bending of perforated circular plates, *International Journal of Mechanical Sciences*,24 (12), 763-775.

[8] Thakur P., Kaur J. & Singh S. B. (2016). Thermal creep transition stresses and strain rates in a circular disc with a shaft having variable density, *Eng. Comput.*, 33(3), 698-712.

[9] Gupta S. K. & Thakur P. (2007). Thermo elastic-plastic transition in a thin rotating disc with inclusion, *Thermal Science*,11(1),103-118.

[10] Seth B. R. (1966). Measure-concept in mechanics, *Int. J Non-linear Mech.*, 1(1), 35-40.

[11] Thakur P., Kaur J. & Sharma P. L. (2024). *New Research on Thermal Stresses*, Nova Science Publisher USA, 1–146.

[12] Thakur P. & Sethi M. (2020). Creep deformation and stress analysis in a transversely material disc subjected to the rigid shaft, *Mathematics and Mechanics of Solids*, 25(1), 17-25.

[13] Sethi M. & Thakur P. (2020). Elastoplastic deformation in an isotropic material disk with shaft subjected to load and variable density, *Journal of Rubber Research*, 23(2), 69-78.

[14] Temesgen A. G., Singh S. B. & Thakur P. (2020). Modelling of creep deformation of a transversely isotropic rotating disc with a shaft having variable density and subjected to a thermal gradient, *Thermal Science and Engineering Progress*, 20 (100745).

[15] Temesgen A. G., Singh S. B. & Thakur P. (2021). Elasto-plastic analysis in functionallygradedthick-walledrotatingtransverselyisotropiccylinderunder a radial temperature gradient and uniform pressure, *Mathematics and Mechanics of Solids*, 26(1),5 -17.

[16] Thakur P., Kumar N. & Sethi M. (2021). Elastic-plastic stresses in a rotating disc of transversely isotropic material fitted with a shaft and subjected to thermal gradient, *Meccanica*, 56(5),1165-1175.

[17] Thakur P., Sethi M., Gupta N., Gupta K. & Bhardwaj R. K. (2021). Thermal stress analysis in a hemispherical shell made of transversely isotropic materials under pressure and thermo-mechanical loads, *ZAMM*, 101(12),e202100208.

[18] Thakur P., Sethi M., Kumar N., Gupta N., Gupta K. & Bhardwaj R. K. (2022). Stress analysis in an isotropic hyperbolic rotating disk fitted with the rigid shaft, *Zeitschriftfür Angewandte Mathematikund Physik.*, 73(1), article id 23,1-11.
[19] Thakur P., Koul M. & Kumar R. (2023). Safety analysis in a thermo-mechanically loaded rotating disk made of polymer fitted with a shaft with variable thickness, *Mechanics of Solids*, 58(3), 818-833.

Chapter 6

Solution of the Boundary Value Problem in Thermoelastic Microelongated Solid by the Finite Difference Method

Praveen Ailawalia[1,*], PhD
Joginder Singh[2], PhD
and Vikas Sharma[2,3], MPhil

[1]Department of Mathematics, University Institute of Sciences, Chandigarh University, Mohali, Punjab, India
[2]Department of Applied Sciences, Chandigarh Group of Institutions, Landran, Mohali-Punjab, India
[3]IK Gujral Punjab Technical University, Kapurthala Punjab, India

Abstract

This chapter employs the finite difference method to solve the two-dimensional problem of a thermo-elastic micro-elongated solid. Explicit solutions are developed for the given problem. The displacement, micro-elongational scalar, and temperature distribution components within the physical domain are determined using explicit numerical methods. All variables are then plotted using MATLAB software. The results are compared, and various outcomes are discussed.

Keywords: thermo elastic, microelongated, displacement, temperature

* Corresponding Author's Email: praveen_2117@rediffmail.com

In: Thermal Modeling Reimagined
Editors: Pankaj Thakur and Jatinder Kaur
ISBN: 979-8-89530-463-1

Introduction

A micro-polar continuum refers to a collection of cohesive particles structured as small rigid bodies undergoing translational and rotational movements. This theory finds practical application in simulating the mechanical behavior of various materials, such as composites reinforced with chopped fibers, platelet composites, aluminum epoxy, and substances like liquid crystals with rigid molecules, rigid suspensions, concrete with sand, and muddy fluids. The increased utilization of these composite materials has spurred significant interest in this theoretical framework.

The theory of linear micro-polar thermo elasticity, as established by Boschi & Iesan [1], explores how each particle within a material can exhibit micro-rotations and volumetric elongation alongside overall bulk deformation. Consequently, a micro-stretch elastic solid is characterized by seven degrees of freedom: three for translation, three for rotation, and one for stretch. Unlike traditional elasticity, the stretching and contracting of material points in micro-stretch bodies are independent of their rotational and translational movements. This generalized solid model provides deeper insights into micro-deformations within material points, making it a valuable mathematical tool for studying diverse media that may not conform to classical elasticity.

The micro-stretch continuum is an effective model in fields such as the study of solids with micro-damage, biological materials like bones, foams, and porous media filled with gas or non-viscous liquids. Materials such as solid-liquid crystals and composites reinforced with chopped elastic fibers can also be categorized as micro-elongated media. Classical theories often fail to accurately represent the behavior of materials with complex internal structures. Eringen & Suhubi [2] and Suhubi & Eringen [3] pioneered nonlinear theories of microplastic solids, leading Eringen [4-7] to develop the linear theory of micro-polar elasticity, which incorporates macro-deformations alongside micro-rotations. His subsequent work introduced a theory of micro-polar elastic solids with axial stretch, expanding the applicability of micro-polar theory to include thermal effects, as discussed by researchers such as Nowacki [8], Eringen [9], Tauchert et al. [10], and Nowacki & Olszak [11]. Thermal effects in micro-polar theory have been extensively explored, highlighting its relevance in understanding the coupled behavior of stress and temperature distributions in thermoelastic media under various sources and boundary conditions. Qalandarov & Khaldjigitov [12] solved mathematical and numerical models of coupled dynamic thermoelastic problems for isotropic bodies. Khaldjigitov et al. [13] developed finite-difference equations for 2D elasticity problems on non-uniform grids. Ailawalia et al. [14] recently solved a boundary value problem in a hygrothermoelastic solid using the finite-difference method.

Overall, these advancements underscore the comprehensive nature of micro-polar theory in addressing the complexities of material behavior, particularly in scenarios involving internal structures and thermal interactions.

Basic Equation

The constitutive equations for a homogeneous, isotropic, micro elongated, thermo elastic solid are given by Shaw & Mukhopadhyay [15]:

$$\sigma_{ij} = \lambda\delta_{kl}u_{r,r} + \mu\left(u_{k,l} + u_{l,k}\right) - \beta_0\left(\frac{\partial}{\partial t}\right)T\delta_{kl} + \lambda_0\delta_{kl}\varphi\,, \quad (1)$$

$$m_k = \lambda_0\varphi_{,k}\,, \quad (2)$$

$$s - \sigma = \lambda_0 u_{k,k} + \mu\left(u_{k,l} + u_{l,k}\right) - \beta_0\left(\frac{\partial}{\partial t}\right)T - \lambda_1\varphi, \quad (3)$$

$$q_k = \frac{K}{T_0}\varphi_{,k}. \quad (4)$$

The field equation of motion, according to Ailawalia et al. [16], Eringen [17], and the heat conduction equation, according to Kiris & Inan [18], for the displacement, micro elongation, and temperature changes are:

$$(\lambda + \mu)u_{j,ij} + \mu\, u_{i,jj} - \beta_0\delta_{2k}\frac{\partial}{\partial t}T_{,i} + \lambda_0\varphi_{,i} = \rho\ddot{u}\,, \quad (5)$$

$$a_0\varphi_{,ii} + \beta_1\frac{\partial}{\partial t}T + \lambda_1\varphi - \lambda_0 u_{j,j} = \frac{1}{2}\rho j_0\,\ddot{\varphi}\,, \quad (6)$$

$$KT_{,ii} - \rho C\frac{\partial}{\partial t}\dot{T} - \beta_0 T_0\frac{\partial}{\partial t}\dot{u}_{k,k} - \beta_1 T_0\dot{\varphi} = 0\,, \quad (7)$$

where $\beta_0 = (3\lambda + 2\mu)\alpha_{t_1}, \beta_1 = (3\lambda + 2\mu)\alpha_{t_2}, \sigma = \sigma_{kk}$ is micro elongation stress tensor, $s = s_{kk}$ is component of stress tensor, δ_{kl} is Kronecker delta, m_k is component of micro stretch vector, λ, μ are lame's elastic constants, $a_0, \lambda_0, \lambda_1$ microelongational constants, C is the specific heat at constant strain, K is the thermal conductivity, α_{t_1} and α_{t_2} are coefficient of linear thermal expansion, ρ is the density of micro elongated medium, j_0 is microinertia, T is the thermodynamic temperature above reference temperature T_0 , φ is microelongational scalar, $\vec{u} = u_i$ is displacement vector. k=1 for Lord-Shulman theory and $k = 2$ for Green-Lindsay theory.

Formulation of the Problem

We have considered a two-dimensional disturbance of medium parallel to *xy*-plane with all physical quantities depending upon (*x, y, t*). For this, we use displacement vector $\vec{u_i} = (u_1, u_2, 0)$. Hence, Equations (5)–(7) become:

$$(\lambda + 2\mu)\frac{\partial^2 u_1}{\partial x^2} + (\lambda + \mu)\frac{\partial^2 u_2}{\partial x \partial y} + \mu\frac{\partial^2 u_1}{\partial y^2} - \beta_0\frac{\partial T}{\partial x} + \lambda_0\frac{\partial \varphi}{\partial x} = \rho\frac{\partial^2 u_1}{\partial t^2}, \quad (8)$$

$$\mu\frac{\partial^2 u_2}{\partial y^2} + (\lambda + \mu)\frac{\partial^2 u_1}{\partial x \partial y} + (\lambda + 2\mu)\frac{\partial^2 u_2}{\partial x^2} - \beta_0\frac{\partial T}{\partial y} + \lambda_0\frac{\partial \varphi}{\partial y} = \rho\frac{\partial^2 u_2}{\partial t^2}, \quad (9)$$

$$a_0\left(\frac{\partial^2}{\partial x^2} + \frac{\partial^2}{\partial y^2}\right)\varphi + \beta_1 T - \lambda_1\varphi - \lambda_0\left(\frac{\partial u_1}{\partial x} + \frac{\partial u_2}{\partial y}\right) = \frac{1}{2}\rho j_0\frac{\partial^2 \varphi}{\partial t^2}, \quad (10)$$

$$K\left(\frac{\partial^2}{\partial x^2} + \frac{\partial^2}{\partial y^2}\right)T - \rho C\frac{\partial T}{\partial t} - \beta_0 T_0\frac{\partial}{\partial t}\left(\frac{\partial u_1}{\partial x} + \frac{\partial u_2}{\partial y}\right) - \beta_1 T_0\frac{\partial \varphi}{\partial t} = 0 \quad (11)$$

The constitutive stress components of the micro-elongation stress tensor are given by the following equations:

$$\sigma_{yy} = \lambda\frac{\partial u_1}{\partial y} + (\lambda + 2\mu)\frac{\partial u_2}{\partial x} - \beta_1 T + \lambda_0\varphi, \quad (12)$$

$$\sigma_{xy} = \lambda\left(\frac{\partial u_1}{\partial y} + \frac{\partial u_2}{\partial x}\right) \quad (13)$$

To simplify calculations, we use the following non-dimensional variables defined by:

$$x' = \frac{\omega^*}{c_1}x,\ y' = \frac{\omega^*}{c_1}y, u_i' = \frac{\omega^*\rho c_1}{\beta_0 T_0}u_i,\ t' = \omega^* t, t_0' = \omega^* t_0, t_1' = \omega^* t_1, \quad (14)$$

$$\sigma_{ij}{}' = \frac{\sigma_{ij}}{\beta_0 T_0}, \varphi' = \frac{\lambda_0}{\beta_0 T_0}\varphi, T' = \frac{T}{T_0} \tag{15}$$

where $\omega^* = \frac{\rho c_1{}^2 C}{\beta_0 T_0}, c_1{}^2 = \frac{(\lambda+2\mu)}{\rho}$.

Using the non-dimensional variables defined by Equation (15) in Equations (8–14), and after dropping the superscripts, we get:

$$\frac{\partial^2 u}{\partial x^2} + a_{11}\frac{\partial^2 v}{\partial x \partial y} + a_{12}\frac{\partial^2 u}{\partial y^2} - \frac{\partial T}{\partial x} + \frac{\partial \varphi}{\partial x} = \frac{\partial^2 u}{\partial t^2} \tag{16}$$

$$a_{12}\frac{\partial^2 v}{\partial x^2} + a_{11}\frac{\partial^2 u}{\partial x \partial y} + \frac{\partial^2 v}{\partial y^2} - \frac{\partial T}{\partial y} + \frac{\partial \varphi}{\partial y} = \frac{\partial^2 v}{\partial t^2} \tag{17}$$

$$\nabla^2\varphi + b_{11}T - b_{12}\varphi - b_{13}\left(\frac{\partial u}{\partial x} + \frac{\partial v}{\partial y}\right) = b_{14}\frac{\partial^2 \varphi}{\partial t^2} \tag{18}$$

$$\nabla^2 T - c_{11}\frac{\partial T}{\partial t} - c_{12}\frac{\partial}{\partial t}\left(\frac{\partial u}{\partial x} + \frac{\partial v}{\partial y}\right) - c_{13}\frac{\partial \varphi}{\partial t} = 0 \tag{19}$$

where,

$a_{11} = \frac{\mu}{\rho c^2}, a_{12} = \frac{\mu}{\rho c^2}, b_{11} = \frac{\beta_1 \lambda_0 c^2}{a_0 \beta_0 \omega^*}, b_{12} = \frac{\lambda_1 c^2}{a_0 \omega^*}, b_{13} = \frac{\lambda_0{}^2}{a_0 \beta_0 \omega^*}, b_{14} = \frac{\rho j_0 c^2}{2a_0}, c_{11} = \frac{\rho C^* c^2}{K^* \omega^*}, c_{12} = \frac{T_0 \beta_0{}^2}{\rho K^* \omega^*}, c_{13} = \frac{\beta_0 \beta_1 T_0 c^2}{\lambda_0 K^* \omega^*}, d_{11} = \frac{\lambda}{\rho c^2}$.

The approximate initial condition is given by:

$$(UX, y, t)|_{t=t_0} = \phi_1, v(x, y, t)|_{t=t_0} = \phi_2, \varphi(x, y, t)|_{t=t_0} = \varphi_0,,$$

$$T(x, y, t)|_{t=t_0} = T_0 \; and \; \frac{\partial u}{\partial t}|_{t=t_0} = \psi_1, \frac{\partial v}{\partial t}|_{t=t_0} = \psi_2 \tag{20}$$

And the boundary conditions are:

$u(x, y, t)|_{x=0} = u_0, u(x, y, t)|_{x=l_1} = \underline{u_0}, u(x, y, t)|_{y=0} = u_0, u(x, y, t)|_{y=l_2} = \underline{u'_0},$

$v(x,y,t)|_{x=0} = v_0, v(x,y,t)|_{x=l_1} = \underline{v_0}\,, v(x,y,t)|_{y=0} = v_0, v(x,y,t)|_{y=l_2} = \underline{v'_0},$

$\varphi(x,y,t)|_{x=0} = \varphi_0, \varphi(x,y,t)|_{x=l_1} = \underline{\varphi_0}\,, \varphi(x,y,t)|_{y=0} = \varphi_0, v(x,y,t)|_{y=l_2} = \underline{\varphi'_0},$

$T(x,y,t)|_{x=0} = T_0, T(x,y,t)|_{x=l_1} = \underline{T_0}\,, T(x,y,t)|_{y=0} = T_0, T(x,y,t)|_{y=l_2} = \underline{T'_0}.$

Finite Difference Equations

We consider a rectangular region bounded by x=0, y=0, $x = l$ and $y = l$(*see* Figure 1). The wave is propagating in the x–y plane. The displacement vector in a two-dimensional slab of silicon material is given as $\vec{u} = (u, v,)$ 0 where $u = u(x, y, t), v = v(x, y, t)$.

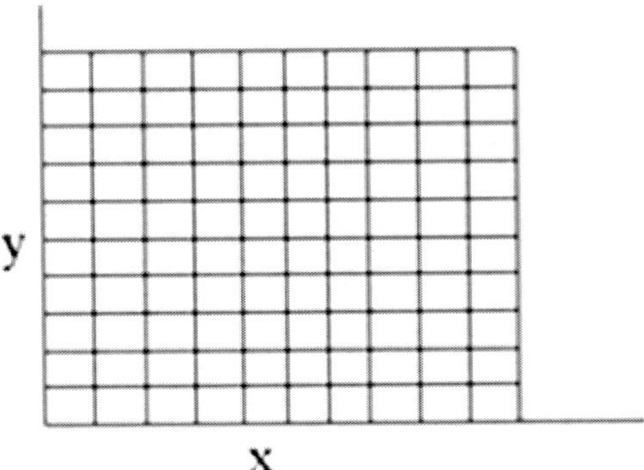

Figure 1. Problem Set up in 2D Region.

We have considered that the equations of motion and coupled Equations of (1)– (3) and constitutive relations (4) in 2-D take form as Rewriting Equations (16-18) as done in using central finite difference approximation using region $t \geq 0$, $0 \leq x \leq l_1$, $0 \leq y \leq l_2$ and boundary points $x = ih_1$, $y = jh_2$, $t = k\tau$ where $i = 0$ *to* n_1, $j = 0$ to n_2 and k = 0, 1, 2, 3 .... Then putting the finite difference approximation respective differential coefficient:

$$u_{i,j}^{k+1} = \tau^2 \left(\frac{u_{i+1,j}^k - 2u_{i,j}^k + u_{i-1,j}^k}{h_1^2} + a_{11} \frac{v_{i+1,j+1}^k - v_{i-1,j+1}^k - v_{i+1,j-1}^k + v_{i-1,j-1}^k}{4h_1 h_2} + a_{12} \frac{u_{i,j+1}^k - 2u_{i,j}^k + u_{i,j-1}^k}{h_2^2} - \frac{T_{i+1,j}^k - T_{i-1,j}^k}{2h_1} - \frac{\varphi_{i+1,j}^k - \varphi_{i-1,j}^k}{2h_1}\right) + 2u_{i,j}^k - u_{i,j}^{k-1} \tag{21}$$

$$v_{i,j}^{k+1} = \tau^2 \left(a_{12} \frac{v_{i+1,j}^k - 2v_{i,j}^k + v_{i-1,j}^k}{h_1^2} + a_{11} \frac{u_{i+1,j+1}^k - u_{i-1,j+1}^k - u_{i+1,j-1}^k + u_{i-1,j-1}^k}{4h_1 h_2} + \frac{v_{i,j+1}^k - 2v_{i,j}^k + v_{i,j-1}^k}{h_2^2} - \frac{T_{i,j+1}^k - T_{i,j-1}^k}{2h_2} + \frac{\varphi_{i,j+1}^k - \varphi_{i,j-1}^k}{2h_2}\right) + 2v_{i,j}^k - v_{i,j}^{k-1} \ \varphi_{i,j}^{k+1} = \frac{\tau^2}{b_{14}} \left(\frac{\varphi_{i+1,j}^k - 2\varphi_{i,j}^k + \varphi_{i-1,j}^k}{h_1^2} + \frac{\varphi_{i,j+1}^k - 2\varphi_{i,j}^k + \varphi_{i,j-1}^k}{h_2^2} + b_{11} T_{i,j}^k - b_{12} \varphi_{i,j}^k - b_{13} \frac{u_{i+1,j}^k - u_{i-1,j}^k}{2h_1} - b_{13} \frac{v_{i,j+1}^k - v_{i,j-1}^k}{2h_2}\right) + 2\varphi_{i,j}^k - \varphi_{i,j}^{k-1} \tag{22}$$

$$T_{i,j}^{k+1} = \frac{\tau}{c_{11}} \left(\frac{()T_{i+1,j}^k - 2T_{i,j}^k + T_{i-1,j}^k}{h_1^2} + \frac{T_{i,j+1}^k - 2T_{i,j}^k + T_{i,j-1}^k}{h_2^2} - c_{12} \frac{u_{i+1,j}^{k+1} - u_{i-1,j}^{k+1} - u_{i+1,j}^{k-1} + u_{i-1,j}^{k-1}}{2h_1 \tau} - c_{12} \frac{v_{i,j+1}^{k+1} - v_{i,j-1}^{k+1} - v_{i,j+1}^{k-1} + v_{i,j-1}^{k-1}}{2h_2 \tau} - c_{13} \frac{\varphi_{i,j}^{k+1} - \varphi_{i,j}^k}{\tau}\right) + T_{i,j}^k \tag{23}$$

By reviewing Equations (20-23), we can determine the values of $u(x,y,t), v(x,y,t), T(x,y,t)$, and $\varphi(x,y,t)$ at the layer t_{k+1}. The values of $u(x,y,t)$, $v(x,y,t)$, $T(x,y,t)$ and $\varphi(x,y,t)$ for the two primary layers $k = 0$ and $k = 1$, can be obtained from the initial conditions for $k=0$:

$$u(i,j) = \phi_1, v(i,j) = \phi_2, T(i,j) = T_0 \ and \ \varphi(i,j) = \varphi_0$$

As we know initially that:

$$\begin{aligned}
&\frac{\partial u}{\partial t}\Big|_{t=t_0} = \psi_1(i,j) \Rightarrow \frac{u_{i,j}^1 - u_{i,j}^{-1}}{2\tau} = \psi_1(i,j),\\
&\Rightarrow u_{i,j}^1 = u_{i,j}^{-1} + 2\tau\psi_1(i,j) \Rightarrow u_{i,j}^{-1} = u_{i,j}^1 - 2\tau\psi_1(i,j),\\
&\frac{\partial v}{\partial t}\Big|_{t=t_0} = \psi_2(i,j) \Rightarrow \frac{v_{i,j}^1 - v_{i,j}^{-1}}{2\tau} = \psi_2(i,j)\\
&\Rightarrow v_{i,j}^1 = v_{i,j}^{-1} + 2\tau\psi_2(i,j) \Rightarrow v_{i,j}^{-1} = v_{i,j}^1 - 2\tau\psi_2(i,j)
\end{aligned} \tag{24}$$

By substituting these values into Equations (20)-(23), we can derive a relationship to calculate the values of $u(x,y,t)$, $v(x,y,t), T(x,y,t)$ *and* $\varphi(x,y,t)$ for $k=1$, giving us the following expression:

$$u_{i,j}^{1} = \frac{\tau^2}{2}\left(\frac{u_{i+1,j}^{0}-2u_{i,j}^{0}+u_{i-1,j}^{0}}{h_1^2} + a_{11}\frac{v_{i+1,j+1}^{0}-v_{i-1,j+1}^{0}-v_{i+1,j-1}^{0}+v_{i-1,j-1}^{0}}{4h_1h_2} + a_{12}\frac{u_{i,j+1}^{0}-2u_{i,j}^{0}+u_{i,j-1}^{0}}{h_2^2} - \frac{T_{i+1,j}^{0}-T_{i-1,j}^{0}}{2h_1} - \frac{\varphi_{i+1,j}^{0}-\varphi_{i-1,j}^{0}}{2h_1}\right) + u_{i,j}^{0} \tag{25}$$

$$v_{i,j}^{1} = \frac{\tau^2}{2}\left(a_{12}\frac{v_{i+1,j}^{0}-2v_{i,j}^{0}+v_{i-1,j}^{0}}{h_1^2} + a_{11}\frac{u_{i+1,j+1}^{0}-u_{i-1,j+1}^{0}-u_{i+1,j-1}^{0}+u_{i-1,j-1}^{0}}{4h_1h_2} + \frac{v_{i,j+1}^{0}-2v_{i,j}^{0}+v_{i,j-1}^{0}}{h_2^2} - \frac{T_{i,j+1}^{0}-T_{i,j-1}^{0}}{2h_2} + \frac{\varphi_{i,j+1}^{0}-\varphi_{i,j-1}^{0}}{2h_2}\right) + v_{i,j}^{0} \tag{26}$$

$$\varphi_{i,j}^{1} = \frac{\tau^2}{2b_{14}}\left(\frac{\varphi_{i+1,j}^{0}-2\varphi_{i,j}^{0}+\varphi_{i-1,j}^{0}}{h_1^2} + \frac{\varphi_{i,j+1}^{0}-2\varphi_{i,j}^{0}+\varphi_{i,j-1}^{0}}{h_2^2} + b_{11}T_{i,j}^{0} - b_{12}\varphi_{i,j}^{0} - b_{13}\frac{u_{i+1,j}^{0}-u_{i-1,j}^{0}}{2h_1} - b_{13}\frac{v_{i,j+1}^{0}-v_{i,j-1}^{0}}{2h_2}\right) + \varphi_{i,j}^{0} \tag{27}$$

$$T_{i,j}^{1} = \frac{\tau}{c_{11}}\left(\frac{T_{i+1,j}^{0}-2T_{i,j}^{0}+T_{i-1,j}^{0}}{h_1^2} + \frac{T_{i,j+1}^{0}-2T_{i,j}^{0}+T_{i,j-1}^{0}}{h_2^2} - c_{12}\frac{u_{i+1,j}^{1}-u_{i-1,j}^{1}-u_{i+1,j}^{0}+u_{i-1,j}^{0}}{2h_1\tau} - c_{12}\frac{v_{i,j+1}^{1}-v_{i,j-1}^{1}-v_{i,j+1}^{0}+v_{i,j-1}^{0}}{2h_2\tau} - c_{13}\frac{\varphi_{i,j}^{1}-\varphi_{i,j}^{0}}{\tau}\right) + T_{i,j}^{0} \tag{28}$$

Numerical Verification

For numerical computations, we consider the values of constants for an aluminum-epoxy-like material as provided by Ailawalia et al. [16], with the corresponding constants, initial conditions, and boundary conditions listed in Table 1. Here, $l_1 = 1$, $l_2 = 1$, at t = 0 , u(x, y, t) = 0, v(x, y, t) = 0, φ (x, y, t) = 0, and T(x, y, t) = T_0Sin(πx/ l_1) Sin (πy/l_2), and boundary conditions are $u(x,y,t), v(x,y,t), T(x,y,t)$ and $\varphi(x,y,t)$, where Γ - boundary of the body. Figure 2 and Figure 3 show the displacement components $u(x,y,t)$ and $v(x,y,t)$. Figure 4 presents microelongation $\varphi(x,y,t)$ and Figure 5 distributions, $T(x,y,t)$ in 3D space at t = 0.09 sec. Every unknown value was calculated using the recurrence with the help of MATLAB software.

Table 1. Physical quantities along with symbol and values

Physical Quantities along with symbol and Values		
Quantity	Symbols	Value with units
Lame's constant	λ	75.9×10^{9} N/m^2
Lame's constant	μ	18.9×10^{9} N/m^2
Density	ρ	21.9×10^{2} Kg/m^3
Temperature	T_0	293 K
Heat Capacity	c	966 JKg^{-1}K^{-1}
Microelongational Constant	a_0	0.61×10^{-10} N
Microinertia	j_0	0.196×10^{-4} m^2
Coefficient of linear thermal expansion	β_0	0.05×10^{5} N/m^{2}K
Coefficient of linear thermal expansion	β_1	0.05×10^{5} Nm^{2}K
Microelongational Constant	λ_0	0.37×10^{10} N/m^2
Microelongational Constant	λ_1	0.37×10^{10} N/m^2
Thermal Conductivity	K	252 J/msK

Table 2. Data tables

Horizontal Displacement Scalar at time t = 0.09										
0	0	0	0	0	0	0	0	0	0	0
0	-0.803	-1.502	-2.067	-2.430	-2.555	-2.430	-2.067	-1.503	-0.803	0
0	-0.715	-1.339	-1.842	-2.166	-2.277	-2.166	-1.843	-1.339	-0.715	0
0	-0.524	-0.981	-1.349	-1.586	-1.668	-1.587	-1.350	-0.981	-0.524	0
0	-0.276	-0.516	-0.710	-0.835	-0.878	-0.835	-0.710	-0.516	-0.276	0
0	0.000	0.000	-0.001	-0.001	-0.001	-0.001	-0.001	0.000	0.000	0
0	0.275	0.515	0.709	0.834	0.877	0.834	0.709	0.516	0.275	0
0	0.523	0.980	1.349	1.586	1.667	1.586	1.349	0.981	0.524	0
0	0.715	1.339	1.842	2.166	2.277	2.166	1.843	1.339	0.715	0
0	0.803	1.502	2.068	2.431	2.556	2.431	2.068	1.503	0.803	0
0	0	0	0	0	0	0	0	0	0	0

Table 3.

Normal Displacement at time t = 0.09										
0	0	0	0	0	0	0	0	0	0	0
0	-0.803	-0.715	-0.524	-0.276	0.000	0.275	0.523	0.715	0.803	0
0	-1.502	-1.339	-0.981	-0.516	0.000	0.515	0.980	1.339	1.502	0
0	-2.067	-1.842	-1.349	-0.710	-0.001	0.709	1.349	1.842	2.068	0
0	-2.430	-2.166	-1.586	-0.835	-0.001	0.834	1.586	2.166	2.431	0
0	-2.555	-2.277	-1.668	-0.878	-0.001	0.877	1.667	2.277	2.556	0
0	-2.430	-2.166	-1.587	-0.835	-0.001	0.834	1.586	2.166	2.431	0
0	-2.067	-1.843	-1.350	-0.710	-0.001	0.709	1.349	1.843	2.068	0
0	-1.503	-1.339	-0.981	-0.516	0.000	0.516	0.981	1.339	1.503	0
0	-0.803	-0.715	-0.524	-0.276	0.000	0.275	0.524	0.715	0.803	0
0	0	0	0	0	0	0	0	0	0	0

Table 4.

Microelongated Scalar at time t = 0.09										
0	0	0	0	0	0	0	0	0	0	0
0	0.0021	0.0039	0.0054	0.0063	0.0066	0.0063	0.0054	0.0039	0.0021	0
0	0.0039	0.0073	0.0101	0.0119	0.0125	0.0119	0.0101	0.0073	0.0039	0
0	0.0054	0.0101	0.0139	0.0164	0.0172	0.0164	0.0139	0.0101	0.0054	0
0	0.0063	0.0119	0.0164	0.0192	0.0202	0.0192	0.0164	0.0119	0.0063	0
0	0.0066	0.0125	0.0172	0.0202	0.0213	0.0202	0.0172	0.0125	0.0066	0
0	0.0063	0.0119	0.0164	0.0192	0.0202	0.0192	0.0164	0.0119	0.0063	0
0	0.0054	0.0101	0.0139	0.0164	0.0172	0.0164	0.0139	0.0101	0.0054	0
0	0.0039	0.0073	0.0101	0.0119	0.0125	0.0119	0.0101	0.0074	0.0039	0
0	0.0021	0.0039	0.0054	0.0063	0.0066	0.0063	0.0054	0.0039	0.0021	0
0	0	0	0	0	0	0	0	0	0	0

Table 5.

Temperature at time t = 0.09										
0	0	0	0	0	0	0	0	0	0	0
0	30.73	54.38	74.62	87.71	92.23	87.73	74.65	54.41	30.74	0
0	54.38	95.93	131.61	154.69	162.67	154.72	131.65	95.98	54.41	0
0	74.62	131.61	180.57	212.24	223.18	212.27	180.63	131.68	74.67	0
0	87.71	154.69	212.24	249.47	262.32	249.50	212.31	154.78	87.77	0
0	92.23	162.67	223.18	262.32	275.84	262.36	223.25	162.75	92.29	0
0	87.73	154.72	212.27	249.50	262.36	249.54	212.34	154.80	87.78	0
0	74.65	131.65	180.63	212.31	223.25	212.34	180.68	131.72	74.69	0
0	54.41	95.98	131.68	154.78	162.75	154.80	131.72	96.03	54.44	0
0	30.74	54.41	74.67	87.77	92.29	87.78	74.69	54.44	30.76	0
0	0	0	0	0	0	0	0	0	0	0

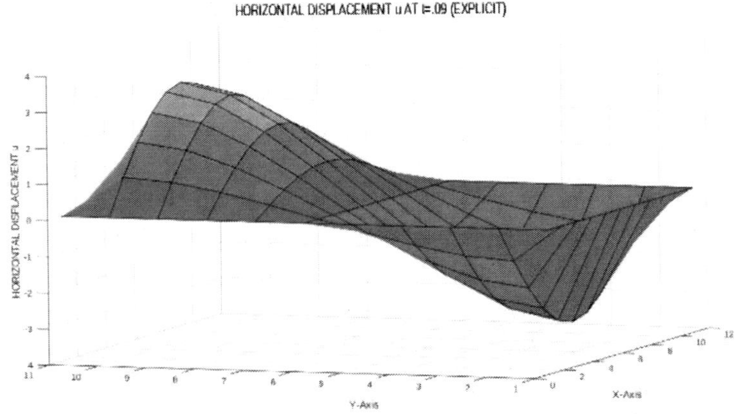

Figure 2. Horizontal Displacement u at t = 0.09.

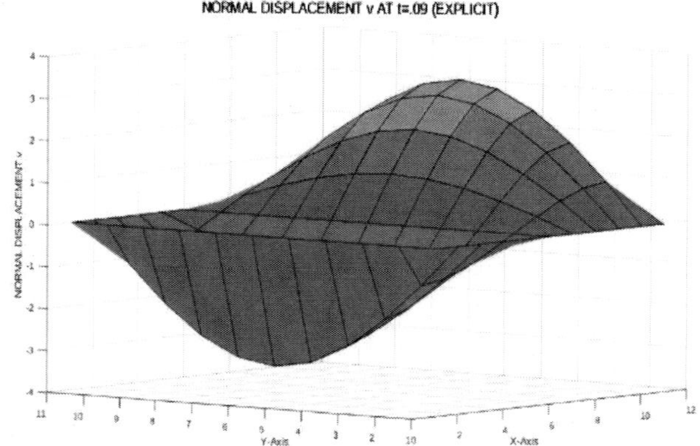

Figure 3. Vertical Displacement v at t = 0.09.

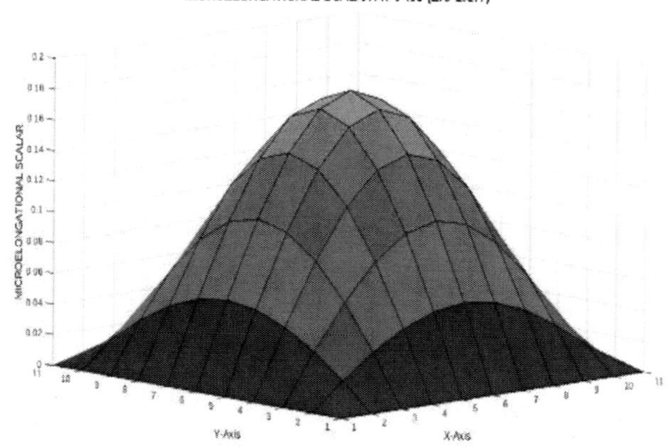

Figure 4. Micro elongation Scalar ϕ at t = 0.09.

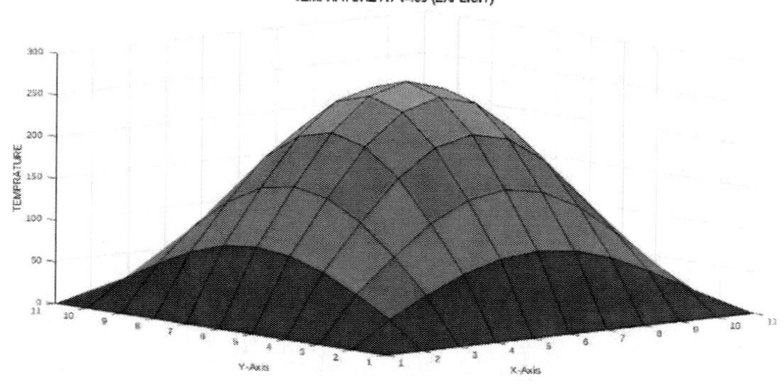

Figure 5. Temperature T at t = 0.09.

Conclusion

We have developed explicit finite difference methods for the two-dimensional coupled thermo elastic micro-elongated solid boundary value problem. The solution was obtained using the recurrence formula algorithm. Numerically obtained solutions show that, with the application of temperature to the solid, the micro-elongation scalar changes in tandem with the applied temperature, as seen in Figures 4 and 5 (or Tables 4 and 5). Changes in horizontal and normal displacement were also recorded and are displayed in Figures 2 and 3 (or Tables 2 and 3), respectively.

Disclaimer

None.

References

[1] Boschi, E. & Iesan, D. (1973). A generalized theory of linear micropolar thermoelasticity, *Meccanica*, 8, 154-157.

[2] Eringen A. C. & Suhubi E. S. (1964), Nonlinear theory of simple micro-elastic solids I, *Int. J. Engineering Science*, 2, 189–203.

[3] Suhubi E. S. & Eringen A. C. (1964) Nonlinear theory of micro-elastic II. *Int. J. Engineering Science*, 2, 389–404.

[4] Eringen A. C. (1965). Linear theory of micropolar elasticity, *ONR Technical Report No. 29*, School of Aeronautics, Aeronautics and Engineering Science, Purdue University.

[5] Eringen A. C. (1966). A unified theory of thermomechanical material, *Int. J. Engineering Science*, 4, 179–202.

[6] Eringen A. C. (1971). Micropolar elastic solids with stretch. In: Prof. Dr. Mustafa InanAnisina, Ari Kitabevi Matbaasi, Istanbul.

[7] Eringen A. C. (1966). Linear theory of Micropolar Elasticity, *J. Mathematics and Mechanics*, 15, 909–923.

[8] Nowacki W. (1966). Couple Stresses in the Theory of Thermoelasticity III, *Bull. Acad. Polon. Sci., Ser. Sci-Tech.*, 14(8), 801–809.

[9] Eringen A. C. (1970). The foundation of micropolar thermoelasticity, *Courses and Lectures, No. 23*, CISM, Udine, Springer-Verlag, Vienna, New York.

[10] Tauchert T. R., Claus W. D., Jr. & Ariman T. (1968). The linear theory of micropolar thermoelasticity, *Int. J. Eng. Sci.*, 6, 36–47.

[11] Nowacki W. & Olszak W. (1974). Micropolar thermoelasticity. In: Micropolar thermoelasticity (W. Nowacki and Olszak (Eds.)), *CISM Courses and Lectures, No. 151*, Udine, Springer-Verlag, Vienna.

[12] Qalandarov A. A. & Khaldjigitov, A. (2020). Mathematical and numerical modeling of the coupled dynamic thermoelastic problems for isotropic bodies, *TWMS Journal of Pure and Applied Mathematics*, 11, 119-126.

[13] Khaldjigitov, A., Kalandarov, A & Djumayozov, U. (2022). Finite-Difference equations for 2D elasticity problems on a non-uniform grid, *AIP Conference Proceedings*, 2637(1), 030005.

[14] Ailawalia, P., Sharma, V & Singh, J. (2024) Finite-difference method for hygrothermoelastic boundary value problem, *Computational Methods for Differential Equations*, 1-13. doi: 10.22034/CMDE.2024.58002.2442.

[15] Shaw S., & Mukhopadhyay B. (2013). Moving heat source response in a thermoelastic microelongated solid, *J. Engineering Physics and Thermophysics*, 86, 716– 722.

[16] Ailawalia P., Sachdeva, S. K. & Pathania D. S. (2016), Internal heat source in thermo elastic microelongated solid at an interface under Green Lindsay theory, *J. Theor. App. Mech.*,46, 65–82.

[17] Eringen A. C. (1999). *Micro continuum Field Theories. I. Foundations and Solids*. Springer Verlag, New York.

[18] Kiris A. & Inan E. (2007) 3D vibration analysis of the rectangular micro damaged plates, *Proc. 8th Int. Conf. on Vibration Problems (ICOVP), India*, 207–214.

Chapter 7

Exploring the Impact of Temperature Variation on Rotating Solid Discs: Angular Speed, Stress and Displacement

Jatinder Kaur[1,*], PhD
Ajay[2], MSc
Abhishek Thakur[3], MSc
and Pankaj Thakur[4], PhD
[1,2,3]Department of Mathematics, Chandigarh University, Mohali, Punjab, India
[4]Department of Mathematics, ICFAI University Himachal Pradesh, Himachal Pradesh, India

Abstract

In this chapter, the radial and circumferential stresses have been computed under the effect of temperature without any externally applied mechanical load. The obtained relations are compared with Gupta's results, assuming no mechanical load. The analysis is further extended to include the effects of heat generation, which helps to understand the influence of thermal stresses on the material's behaviour.

Keywords: temperature, disc, stress, yield, displacement

* Corresponding Author's Email: jksaini83@gmail.com

In: Thermal Modeling Reimagined
Editors: Pankaj Thakur and Jatinder Kaur
ISBN: 979-8-89530-463-1

Introduction

When exposed to heat, the thermal effects on a solid disc are diverse, primarily from the material's expansion as its temperature rises, leading to thermal expansion, dimensional alterations, stress and strain, warping and buckling, and thermal stress relief. As the solid disc undergoes thermal expansion due to heating, the increased temperature prompts more vigorous vibration among its atoms or molecules, which increases the average separation between them, resulting in the disc's expansion. The degree of this expansion is contingent upon the material's coefficient of thermal expansion, a property unique to each material [1, 2]. This expansion can induce internal stresses within the material, stemming from the constraints imposed by its surroundings or other factors on the material's expansion. The magnitude and distribution of these stresses depend on the disc's geometry, boundary conditions, and material properties. Upon cooling after being heated, solid disc contracts, potentially generating residual stresses within the material. This can be alleviated through thermal stress relief methods such as annealing or controlled cooling. These methods aim to reduce the risk of dimensional instability or failure [3-5].

The material's behaviour has been scrutinized concerning linear strain hardening, yield criteria, and the associated flow rule, with stress distribution determined by aligning plastic stresses at corresponding disc radii, as Gamer discussed [6, 7]. Dastidar & Ghosh [8] investigated the temperature-induced stress distribution in a solid sphere resulting from a constant heat source, using the Ramberg-Osgood stress-strain relation, a widely used empirical equation describing the nonlinear stress-strain behaviour of materials under uni-axial loading conditions. This Equation is commonly employed in engineering and materials science to characterize the deformation of metals, alloys, and other materials exhibiting plasticity and significant strain hardening. Ghosh [9] examined the impact of temperature on the transverse vibration of a rotating disc of varying thickness, considering the presence of a consistent heat source. Guven & Altay [10] explored how non-uniform heat generation influences plastic deformation and stress distribution in a material, particularly under specific stress conditions. You & Zhang [11] conducted a study on the analysis of elastic-plastic stresses in a rotating solid disc, presenting numerical examples and comparing the results with finite element analysis, which showed similar outcomes.

Additionally, Seth [12, 13] introduced a transition theory as an alternative approach that does not require assumptions like yield conditions or incompressibility. This theory addresses a more general problem and allows for the derivation of specific cases based on desired assumptions. The spectrum of citations from [14-17] demonstrates the integration of asymptotic solutions at critical or turning points of the governing differential equations, successfully applying the theory to various engineering situations.

This chapter delves into the study of thermo elastic-plastic stresses induced by heat on a rotating solid disc, with graphical representations illustrating the impact of temperature. In general, spatial or temperature factors may influence steady-state temperature and heat generation, as noted by Thakur [18]. This paper focuses on elastic-plastic stress analysis in the finite deformation of a solid disc under the influence of heat generation. The Equation describing how the heat generation rate varies radially within the solid disc, where q˙(r) is determined by the radial position, can be expressed as:

$$\dot{q}(r) = q_0\left[1-\left(\frac{r}{a}\right)^s\right] \tag{1}$$

Basic Equations

Seth discussed the strain components for infinitesimal deformation using cylindrical polar coordinates [13]. In the developed model, the object under consideration is a disc of radius a with uniform density, implying that the density throughout the disc remains constant and is denoted by ρ. The angular velocity ω measures how fast the disc is rotating about its axis, with the axis of rotation perpendicular to its plane. The standard Equation relating stress and strain for isotropic materials is provided by Sokolnikoff [5]:

$$\tau_{ij} = 2\mu e_{ij} + \lambda\delta_{ij}e_{kk} - \xi T\delta_{ij} \tag{2}$$

where T is temperature and $i, j = 1, 2$ and 3 respectively.

The mathematical representation of one-dimensional heat conduction in a steady state with a heat source can be found in reference [16]:

$$\frac{1}{r}\frac{d}{dr}\left[r\frac{dT(r)}{dr}\right] + \frac{\dot{q}(r)}{k} = 0 \text{ for all } 0 < r < a. \tag{3}$$

The temperature field impeccably meets Equation (3). Also $\frac{dT(r)}{dr}=0$ at $r = 0$ and $T(r) = 0$ at $r = a$. Employing the prescribed boundary conditions alongside Equation (1), we drive the steady-state temperature distribution in the following manner:

$$T(r)=\frac{q_0 a^2}{4k}\left[1-\left(\frac{r}{a}\right)^2-\frac{4}{(s+2)^2}\left[1-\left(\frac{r}{a}\right)^{s+2}\right]\right]$$

The heat conductivity, represented by k, leads to the transformation of the problem's expression into Equation (2):

$$\tau_{rr}=\lambda I_1 I_2+2\mu e_{rr}-I_2\xi T\,,\,\tau_{\theta\theta}=\lambda I_1 I_2+2\mu e_{\theta\theta}-I_2\xi T\,,$$
$$\tau_{\theta z}=\tau_{zz}=\tau_{r\theta}=\tau_{zr}=0 \tag{4}$$

where $I_1=\frac{2\mu}{\lambda+2\mu}$ and $I_2=e_{rr}+e_{\theta\theta}$.

The derivation of strain components in relation to stresses, as given by Equation (2), stems directly from the equation and is expressed as:

$$e_r=\frac{\partial u}{\partial r}=I_3\left[\tau_{rr}-\nu\,\tau_{\theta\theta}\right]+\alpha T\,,\,e_\theta=\frac{u}{r}=I_3\left[\tau_{\theta\theta}-\nu\,\tau_{rr}\right]+\alpha T\,,$$
$$e_z=-\nu I_3\left[\tau_{rr}-\tau_{\theta\theta}\right]+\alpha T\,,\,e_{rr}=e_{\theta z}=e_{zr}=0 \tag{5}$$

where $I_3=1/E$. The generalized components of stress and strain in the radial, tangential, and axial directions are defined by Thakur et al. [16]. All equilibrium equations are satisfied, except for one:

$$\frac{d}{dr}(rT_{rr})-T_{\theta\theta}+\rho\omega^2 r^2=0 \tag{6}$$

Utilizing the stress components as defined in Thakur et al. [16] and Equation (3) in Equation (6), we derive a nonlinear differential equation in β:

$$\frac{dP}{d\beta} = \frac{1}{n\beta^2 P(2-C)\{2\beta P+2\beta-1\}^{\left(\frac{n}{2}-1\right)}}\left[\left\{\frac{n\rho\omega^2 r^2}{2\mu} - \{2\beta P - 2\beta - 1\}^{\frac{n}{2}}\left[1 + \frac{(n\beta P2+n\beta P)(2-C)}{\{2\beta(P+1)-1\}}\right] + \{2\beta - 1\}^{\frac{n}{2}}\left[1 - \frac{n\beta P(1-C)}{\{2\beta-1\}}\right]\right\} - \frac{nC\xi q_0 \mathrm{r}}{8\mu k}\left[\frac{4}{(s+2)}\left(\frac{r}{a}\right)^s - 2\right]\right] \tag{7}$$

where $P = f(\beta)$ and $T'(r) = \frac{q_0 r}{4k}\left[\frac{4}{(s+2)}\left(\frac{r}{a}\right)^s - 2\right]$.

The transition points delineating β in Equation (7) are characterized by $P = -1$ and $P = \infty$.

Boundary conditions:

$$u = 0 \; at \; r = 0 \text{ and } T_{rr} = 0 \; at \; r = a \tag{8}$$

Evaluation of the Stresses of the Problem

Critical to evaluating the plastic stress, it is imperative to establish the transition function R_2 in the crucial transition point $P \to \pm\infty$, a methodology delineated by [12-18], and the transition function obtained as follows:

$$R_2 = A_2 r^{-1/(2-C)} \tag{9}$$

By applying the boundary conditions, the constant of integration A_2 can be precisely determined. Synthesizing Equations (9), we unveil the stresses and displacement relations as follows:

$$\begin{aligned} rT_{rr} &= \frac{\rho\omega^2}{9}\left(r^3 - a^3\right)\frac{(1-C)(5-4C)}{(3-2C)} - \rho\omega^2\left(r^3 - a^3\right) \\ &- \frac{\alpha E q_0 a^3 (2-C)^2}{k(s+2)(s+3)^2(3-2C)}\left\{\left(\frac{r}{a}\right)^{s+3} - 1\right\} - \frac{\alpha E q_0 a^2 (1-C)^3 r}{2k(3-2C)}\left\{\left(\frac{r}{a}\right)^2 - 1\right\} \\ &+ \frac{(2-C)(s+4)\alpha E q_0 a^3 (1-C)^3}{k(3-2C)(s+3)(s+2)^2}\left\{\left(\frac{r}{a}\right)^{s+3} - 1\right\} - \frac{\alpha E q_0 a^2 (2-C) r}{4k}\left\{1 - \left(\frac{r}{a}\right)^2\right\} \\ &+ \frac{\alpha E q_0 a^3 (2-C)}{k(s+2)(s+3)}\left\{1 - \left(\frac{r}{a}\right)\frac{(s+3)}{s+2} + \frac{1}{(s+2)}\left(\frac{r}{a}\right)^{s+3}\right\} \end{aligned} \tag{10}$$

$$rT_{\theta\theta} = \frac{\rho\omega^2}{9}\left(r^3 - a^3\right)\frac{(1-C)^2(5-4C)}{(3-2C)(2-C)} - \frac{\rho\omega^2 a^3(1-C)}{3(2-C)}$$
$$- \frac{\alpha E q_0 a^3(1-C)(2-C)}{k(s+2)(s+3)^2(3-2C)}\left\{\left(\frac{r}{a}\right)^{s+3} - 1\right\} - \frac{\alpha E q_0 a^2(1-C)^4 r}{2k(3-2C)(2-C)}\left\{\left(\frac{r}{a}\right)^2 - 1\right\}$$
$$+ \frac{\alpha E q_0 a^3(1-C)^4(s+4)}{k(s+2)^2(s+3)(2-C)(3-2C)}\left\{\left(\frac{r}{a}\right)^{s+2} - 1\right\} - \frac{\alpha E q_0 a^2(2-C)r}{k(s+2)^2}\left\{1 - \left(\frac{r}{a}\right)^2\right\} + \frac{\alpha E q_0 a^3(1-C)}{k(s+2)(s+3)} \tag{11}$$

$$u = \frac{I_3(1-C)}{(2-C)}\left[\frac{\rho\omega^2 r^3}{3} - \frac{\alpha E q_0 a^2(1-C)r}{2k}\left\{-1 + \left(\frac{r}{a}\right)^2 - \frac{2(s+4)}{(s+2)^2(s+3)}\left(\frac{r}{a}\right)^{s+2}\right\}\right] \tag{12}$$

These results align with those obtained by Kaur et al. [14] when the mechanical load is set to zero. Numerical calculations reveal that the obtained stresses and displacement values when compared with those under the absence of mechanical load, demonstrate distinct patterns. Specifically, circumferential stresses peak on the inner surface of a rotating disc composed of compressible material, contrasting with radial stresses, which peak on the outer surface—a phenomenon not observed in discs made of incompressible materials. Moreover, incorporating thermal effects in compressible materials intensifies circumferential stress on the inner surface and radial stress on the outer surface relative to their incompressible counterparts. Notably, thermal effects exacerbate both circumferential and radial stresses on both internal and external surfaces in the fully plastic state.

Conclusion

The results indicate that rotating discs made of compressible materials require a higher angular velocity to penetrate the inner surface compared to discs made of incompressible materials. Additionally, as the radius ratio increases, a substantially higher angular velocity is required for yielding. Moreover, including thermal effects in the plate accentuates the circumferential stress on the inner surface and radial stress on the outer surface, particularly for compressible materials, in contrast to their incompressible counterparts. This finding underscores the intricate interplay between material properties and thermal effects on the mechanical behaviour of rotating discs, offering valuable insights for design and performance optimization.

References

[1] Timoshenko, S. P. & Goodier, J. N. (1970). *Theory of Elasticity*, Mc Graw-Hill, New York.

[2] Blazynski T. N. (1983). *Applied Elasto-plasticity of Solids*, McMillan Press Ltd., London.

[3] Johnson W. &Mellor P. B. (1962). *Plasticity for Mechanical Engineers*, Van-Nostrand Reinhold Company, London.

[4] Chakrabarty J. (1987). *Theory of Plasticity*, McGraw-Hill, New York.

[5] Sokolnikoff I. S. (1952). *Mathematical theory of Elasticity*, 2nd edition, New York, 65-79.

[6] Gamer U. (1984). Elastic-plastic deformation of the rotating solid disk, *Ingenieur Archiv*, 54, 345-354.

[7] Gamer U. (1985). Stress distribution in the rotating elastic-plastic disk, *ZAMM*, 65, T136-137.

[8] Dastidar D. G. & Ghosh P. (1972). Temperature stress distribution in a sphere due to uniform heat source for strain hardening material, *Journal of Applied Mechanics*, 39, 274-276.

[9] Ghosh N. C. (1975). Thermal effect on the transverse vibration of spinning disk of variable thickness, *Journal of Applied Mechanics*, 42, 358-362.

[10] Guven U. & Altay O. (2000). Elastic-plastic solid disk with non-uniform heat source subjected to external pressure, *International Journal of Mechanical Sciences*, 42, 831-842.

[11] You L. H. &Zhang J. J. (1999). Elastic-plastic stresses in a rotating solid disk, *International Journal of Mechanical Sciences*, 41, 269–282.

[12] Seth B. R. (1962). Transition theory of Elastic-plastic deformation, Creep and relaxation, *Nature*, 195,896 -897.

[13] Seth B. R. (1966). Measure concept in Mechanics, *Int. J. Nonlinear Mech.*,1, 35- 40.

[14] Kaur J., Thakur P. & Singh S. B. (2016). Steady thermal stresses in a thin rotating disc of finitesimal deformation with mechanical load, *Journal of Solid Mechanics,* 8, 204-211.

[15] Gupta S. K. & Thakur P. (2008). Creep Transition in an isotropic disc having variable thickness subjected to internal pressure, *Proceeding National Academy of Science, India, Springer*, Sec. - A, 78, 57-66.

[16] Thakur P., Singh S. B. & Kaur J. (2014). Elastic-plastic stresses in a thin rotating disk with a shafthaving density variation parameter under steady-state temperature, *Kragujevac J. Sci*, 36, 5-17.

[17] Thakur, P. (2012). Stresses in a thin rotating disc of variable thickness with rigid shaft, *Advanced Technologies and Materials*, 37(1), 1–14.

[18] Thakur P. (2014). Steady thermal stress and strain rates in a rotating circular cylinder under steady-state temperature, *Thermal Science*, 18, 93-106.

Index

S

T

U

V

Y